Firefighting with Henry's Model T

William D. Killen

William D. Killen, Publisher
526 Whitetail Road
Church Hill, TN 34762

1st. Edition, 2008

ISBN 978-0-615-22303-2

Printed in the United States of America

Cover design by Elizabeth Brockman

Edited by William & Carole Killen

COVER CREDIT: 1919 Ford Model T Howe pumper, PHOTO CREDIT: Mark Radtke'; Ex-Lille, Maine fire department's 1927 Ford Model T Boyer Chemical Chief's car, PHOTO CREDIT: Tom Herman; and Brigden, Ontario's 1926 Ford Model T Gauld pumper. PHOTO CREDIT: Herb Galloway

BACK COVER: 1909-10 Ford Model T Fire Chiefs Car PHOTO CREDIT: Walt McCall Collection; 1921 Waterous Ford Model T Pumper PHOTO CREDIT: Walt Schryver ca 1975; 1922 Ford Model TT Brockway PHOTO CREDIT: Russell Dawson

Dedication

This book is dedicated to my lovely wife Carole—my best friend and helpmate who has made my life and career an enjoyable success, blessed me with four beautiful children and tolerated my fire service activities and hobbies for nearly fifty years.

Also to the Model T Ford Club of America, the Model T Ford Club International and the Society for the Preservation and Appreciation of Antique Motor Fire Apparatus in America, three great organizations that have contributed to the preservation of the Model T Ford and fire fighting apparatus, as well as the history of these wonderful vehicles for future generations.

Acknowledgments

One of the first things I noticed after moving to Tennessee was that East Tennesseans often expressed their appreciation with the words ***"I appreciate you"*** rather than the usual "Thank You".

To each and everyone who has assisted me in this project, ***"I appreciate you!"***

There were several individuals who shared materials from their personal collections and put forth an extra effort to help with this project, a sincere ***"I appreciate you"*** Steve Hagy, Tom Herman, Matt Lee, Walt McCall and Tod Wirth. The comments, information, photographs, advertising literature, and recommendations each of you shared were quality additions that will ensure the success of this book.

Also, a special ***"I appreciate you!"*** to:

Dr. Will Brooks	Liz Brockman	Jim Golden	Phil McLean
Grant Mishoe	John Purkey III	Bob Rohrback	Joel Woods.

Other contributors:

Randy Acker, Jim Atkinson, Ex-Chief Lester Arsell, Lester Beachy, Blain Bechtold, Tim Boaden, Leon Brewer, W. Parker Browne, Ken Buohl, Chief Mike Chiaramonte, Peter Darley, William Darrow, Chief Russell Dawson, Bob Dubbert, Battalion Chief Paul Frazier, Chief Bill Friedrich, Herb Galloway, Phil Gilson, Dan Jasina, Chief Dan Jones, Gary Kadzielawski, Michael Kitsock, Chief Jim Lamotte, Dennis Maag, Scott Mattson, Wes Melo, Malden Miller, Howard Miele, Chief Don Newman, Dean Nickerson, Neil and Mary Ann O'Brien, Chief Terry Ott, Ed Peterson, Ron Czaplicki, John Pollard, Fireman's Fund Insurance Company; Mark Radtke, Chief Bob Rebtoy, Ray Richardson, Glenn Robinson, Deputy Chief Mike Robitaille, Chief Paul Romano, Nick Sage, Chief Bill Sheehy, Stuart Warner, Scott Wejmar, Captain Brian White, Ken Wilson, Joe Woyjeck and Jim Young.

Foreword

It has been acknowledged by many people around the world that no invention altered the lifestyle of humanity as much as the automobile and the Ford Model T are head and shoulders above the other marques of that era. At one point in time more than half the automobiles and trucks on the road around the world were Model T Fords.

Not only did the automobile and truck alter the lifestyle of humanity, it altered tactics, improved mobility and logistics for the military, and became a major factor in reducing response time to fires and influencing firefighting tactics.

Many of the first manufacturers of automobiles and trucks were wagon and carriage makers and during World War I several wagon makers such as Stoughton Wagon Works, Stoughton, WI. built "Liberty Trucks" for the Army.

At the beginning of the twentieth century communities and towns across the country fought fires with horse drawn steam pumpers and hand pulled hose carts and chemical wagons. Henry Ford's Model T, in a very big way, changed the way fire departments responded to fires and improved the delivery of extinguishing agents in controlling fires. Adding small chemical tanks to the Fire Chief's Ford Model T Runabout or removing the forward portion of horse drawn motorized fire pumps and mounting the pump on the Model T Ford chassis ensured a quicker response of men and equipment to the fire.

Bruce McCalley's book, Model T Ford, The Car that Changed the World, defines the changes in the Model T from the beginning of production in 1908 until the last model rolled off the line in 1927. Model T enthusiasts consider McCalley's book to be the bible on the Model T Ford. Ford enthusiasts around the world will celebrate the 100th Anniversary of the Model T this year (2008).

My book, Navy and Marine Corps Fire Apparatus: 1836-2000 Photo Gallery chronicled the history of firefighting apparatus used at Navy and Marine Corps installations around the world. I was not able to identify the year and manufacturer of the first Model T fire apparatus used by the Navy, but the earliest documented in the book was a 1919 Model Ford TT 1-Ton with a Peter Pirsch Chemical system. The Army purchased hundreds of American LaFrance and Howe Model T pumpers and hose wagons before World War I and positioned these trucks at Army camps throughout the United States.

This pictorial history of fire chief cars and firefighting vehicles built on Ford Model T and TT chassis may serve to provide a little more information and a different perspective on the history of the Model T. Everything the reader might want to know about the "nuts and bolts" of the Model T's in this pictorial history can be found in Bruce McCalley's book.

As I gathered material for this book many interesting stories came with the photographs and history of the vehicles. I included several of these stories about individual Model T fire engines that were important elements in community fire service long after the units were taken out of service and the people whose lives made a difference in their communities with the Model T Ford.

The story of Captain John Purkey, Jr. of the Violetville Volunteer Fire Department in Baltimore County, Maryland, "a man who rode it, drove it, restored it and was buried off it" or the early morning fire in New Jersey that destroyed Nesco VFD's 1925 fire engine and the small blaze in the flat above the New Washington, Ohio fire chief's hardware store in December, 1920 are three interesting examples.

Introduction

The Model T Ford was officially introduced to the world in October 1908 and remained in production until May 1927. There were over 15,000,000 vehicles produced during that period and the Ford Motor Company continued to build engines and some parts for that vehicle until 1941. That production record held for many years and the impact that all those automobiles made in America has lasted even longer.

There are hundreds of books that have been written about the Model T Ford and just as many books have been written about fire trucks and the fire service, since 1908, when that auto was introduced. None of those books were specifically written to address or recognize the contributions that the Model T Ford made to the fire service and the numerous fire apparatus companies that produced these vehicles.

There are a few fire companies in the United States that still have their original Model T Fire Chief's car or truck and have kept the vehicle in good repair for parades and promotions. Some have actually bought their original vehicles back or restored their old stock that was kept, at a considerable expense. Bill Killen attempts to specifically locate, identify and document the Ford Model T Fire Equipment that remains and credit those fire departments, associations and individuals responsible for saving these historic vehicles.

Chief Bill's interest in fire equipment started about the time he heard his first fire siren, continued as a teenage volunteer, escalated to a full time paid firefighter position and continued into fire service management as a fire chief and Director, Navy Fire and Emergency Services. His first interest in the Model T Ford goes back over 40 years, but that vehicle he purchased was owned very briefly at that time. He longed for another Model T fire car for many years and finally purchased one of his favorites. This car is in excellent condition and was restored by his brother-in-law, Marty Scott, who is also a retired Navy firefighter.

Bill has collected photographs, slides, operator's manuals, and advertisements of fire apparatus for about fifty years, but his decision to publish a book on Model T Ford fire apparatus is recent and primarily prompted by his purchase of his 1925 Ford Fire Chief's Car.

He has traveled extensively with his work, in the U. S. Navy Fire Service, and during his term as the President of the International Association of Fire Chiefs 2005-2006. He has made many friends and most have offered a photo or story or two that have been included in this book. This is Bill's third book and a few of these photos were previously published in First to the Flames, The History of Fire Chief Vehicles, by Edward L. Peterson (Past President of the Society for the Preservation and Appreciation of Antique Motorized Fire Apparatus in America), ISBN 0-87341-674-0 and his "Navy & Marine Corps fire Apparatus, 1836 – 2000 Photo Gallery", ISBN 1-58388-031-3. His recent Internet contacts have also added several of these photos.

Jim Golden
Model T Enthusiast
Bowie, Maryland

There is something magical about a fire engine and the mystique about the heroic deeds and stories of firefighters. Whether it was the sound of horses hooves on cobble stone streets and clanging bells in the 1800's, or the roar of sirens on open cab fire trucks, firefighters hanging on to the tailboard in the 1950's, or the modern day sound of electronic sirens and diesel powered fire trucks responding through our communities today, everyone, young and old seems to enjoy the excitement a fire truck brings. PHOTO CREDIT: Steve Hagy Collection

It is not surprising that the awe and excitement of fire trucks is not limited to children and not everyone who owns a toy fire truck is a little boy or girl. Multiple surveys taken the past thirty years document that firefighters are the most trusted and respected people in North America. The interest in owning an antique fire truck or fire chief's car or a firefighter's helmet is a hobby enjoyed by thousands of people around the world and may very well be influenced by the admiration, respect and trust people have in their firefighters.

The interest in firefighting can be traced as far back as the Romans when an organization known as "Vigils" was formed by Emperor Augustus after a disastrous fire in 6 A. D. The Vigils lasted for more than 500 years and at the height of power in the Roman Empire numbered more than 16,000. These "Vigils" would patrol the city at night carrying buckets, an axe, and a hook (i.e. a pike.

Collection of various styles of wooden rattles

Early colonial history records that church bells called the community to a fire and a watchman patrolled colonial communities at night to watch for the start of fires. In the event a fire occurred, a wood rattle was used to alarm the sleeping community.

Authentic quality reproduction of Colonial Era leather fire buckets by Steven Lalioff.

The watchman would give the alarm, shouting “Throw out your buckets”. Every homeowner was required to own two leather fire buckets and threw them into the street where lines were formed to fill the buckets with water from cisterns and creeks and pass them down the line where the water was thrown on the fire.

The first American firefighters were ordinary citizens ready to respond when the bell rang. After the great Boston fire in 1760, Bostonians petitioned King George for help and fumed at his indifference to their fire problem. Patriot leaders who also served in fighting fire included Sam Adams, Paul Revere, Ben Franklin, John Hancock and George Washington.

Fire fighting apparatus evolved from hand held buckets to hand operated pumps, which were later mounted on hand drawn carts such as this Rumsey #1, hand drawn fire pumper, circa 1875-1876. PHOTO CREDIT: Northeast Auctions.

Hose carts and pumps were used for many years until the development of the steam fire engine nearly two hundred years ago in England. Many fire departments maintained steam engines in service for several years after World War One

Horse drawn hose wagon circa 1915

Horse drawn fire apparatus appeared and continued in active use well into the late 1920's. The use of horse drawn fire vehicles began in the mid 1860's and consisted of steam engines, hose carts and hose reels, hook and ladder rigs and the fire chief's buggy. A number of horse or hand drawn fire apparatus were fitted with gasoline powered fire pumps by manufacturers such as American LaFrance and Howe Fire Apparatus companies, among others. PHOTO CREDIT: Bill Killen Collection

I really do not know as much about Model T Ford's as I would like and I want the reader to clearly understand that this book IS NOT an authoritative text on the Model T Ford. I truly am a novice when it comes to the Model T. Whenever I have a question or need to research something about the Model T, I reach for my copy of Bruce McCalley's book, "Model T Ford" The car that Changed the World or call my friend, "Mr. Friendly Fix-It", Jim Golden or visit the Model T Forum on the web.

However, I do know a little something about the fire service and firefighting vehicles, having written specifications, developed fire apparatus requirements and coordinated procurement of more than $100 million in fire apparatus. I have visited the manufacturing facilities of American LaFrance, Emergency One, Pierce, Oshkosh, Seagrave, FWD, Sutphen, Ward LaFrance, Maxim, Peter Pirsch, Fire Tec, KME, LTI, Young, Hale, Waterous, 3D, and Ward 79, just to name a few.

This pictorial history is intended to explore through pictures and advertisements the contributions and impact the Model T Ford had on the delivery of firefighting services in communities, large and small, all across North America. I believe many readers will respond with additional pictures and information about Model T Ford fire apparatus that should have been included in this history, I certainly hope so, maybe there will be enough additional material for a second volume.

AMERICAN-LAFRANCE CHEMICAL
FOR SMALL TOWNS AND VILLAGES

ON STANDARD FORD CHASSIS

Fire apparatus manufacturers in the United States and Canada who built fire apparatus on the Model T Ford chassis included Ajax Fire Engine Works, Brooklyn, NY, American LaFrance Fire Engine Company, Elmira, NY, Bickle Fire Engine Company, Ontario, Canada, Boyer Fire Apparatus Company, Logansport, IN, Buffalo Fire Appliance Company, Buffalo, NY, O. J. Childs Company, Utica, NY, Fire Equipment Limited, Montreal, Canada, Foamite Childs Corporation, Utica, NY, W.S.Darley & Company, Chicago, IL, Hale Fire Pump Company, Conshohocken, PA, Hallock Engineering Company, Medina, OH, O. N. Hirst Company, Placerville, CA, Howe Engine Company, Indianapolis, IN, Howe Fire Apparatus Company, Anderson, IN, Lorne Tractor / Fire Engine Company, Ingersoll, Ontario, Canada; Luverne Fire Apparatus Company, Luverne, MN, D. E. McCann's Sons, Portland, OR, Obenchain-Boyer Company, Logansport, IN, Julius Pearse, Denver, CO, Peter Pirsch Company and Peter Pirsch & Sons, Kenosha, WI, Prospect Manufacturing Company, and Prospect Fire Engine Company, Prospect, OH, Southern Fire Apparatus Company, Dallas, TX C. H. Sutphen & Son, Columbus, OH, Pierre Thibault Ltd, Quebec, Canada and The Waterous Company, ST. Paul, MN.

In addition to these manufacturers many cities, towns and fire departments added the Smith Form-a-Truck device and added a chain-driven transmission to the Model T to construct their own fire engine. Many of these locally constructed fire vehicles survive today.

The American LaFrance Factory at Elmira, New York

The American LaFrance Corporation filed for Chapter 11 bankruptcy in January, 2008. One of the oldest names in fire apparatus manufacturers, American LaFrance can trace its roots back to 1832. PHOTO CREDIT: Steve Hagy Collection

1909 Ford Model T Fire Chiefs Car

This beautiful 1909 Model T Fire Chief's car owned by Pierre Suprenant was photographed at the 2004 Montreal Fire Department's gala parade of antique and modern fire apparatus. PHOTO CREDIT: Mark Bindham

1909 Ford Model T Piston Pumper

Brookline, New Hampshire's 1909 Model T Ford with a Piston Pump, manufacturer unknown. The headlights on this rig do not appear to be correct for that year. PHOTO CREDIT: Collection of Thomas L. Herman

1909-10 Ford Model T Fire Chiefs Car

This may be a Canadian-built Ford, similar to but not the same as cars produced in the U.S. Ford made the right hand drive for some provinces of Canada and for export to the British Commonwealth. PHOTO CREDIT: Walt McCall Collection

Hamburg Volunteer Fire Department's 1912 Ford Model T at Station No. 1

Hamburg VFD's first motorized apparatus; a Ford Motel T chemical was purchased in 1912, and later sold to the Scranton Fire Company. This truck is still in the possession of the Scranton Fire Company and was recently restored. PHOTO CREDIT: Village of Hamburg, New York

1911 Ford Model T Fire Chiefs Car

Bill Ackerson treats son Charlie to a ride in his 1911 Fire Chiefs car in the 2008 Mahwah, New Jersey Memorial Day Parade. PHOTO CREDIT: William Ackerson Collection

1913 Ford Model T American LaFrance Double Tank Chemical

Ron Czaplicki restored this 1913 American LaFrance double tank chemical in 1980 and later sold the truck to a collector in Germany. PHOTO CREDIT: Ron Czaplicki

1913 Ford Model T American LaFrance

Campbell (CA) Fire Department's First Fire Truck, with two chemical tanks, 12' (4m) ladder, two oil lanterns, two axes and 50' (18m) of 1" (25mm) rubber hose. The Ford American LaFrance double tank chemical car with hose trailer was purchased from the American La France Fire Engine Company, Inc., in Elmira, NY, at a cost of $1775. PHOTO CREDIT: Campbell Historical Museum.

The American-La France Double Tank Chemical and Hose Car with Trailer

This is one of several pictures in a 1916 American LaFrance Fire Engine Company brochure titled "Chemical Fire Apparatus mounted on Ford Chassis" distributed to dealers and fire departments throughout the United States and Canada. This advertisement is similar to the unit delivered to the Campbell, California fire department in 1913. PHOTO CREDIT: Steve Hagy Collection

1913 Ford Model T O. J. Childs Double Tank Chemical

This 1913 Ford Model T is one of the earliest known Double Tank Chemicals built on the Ford Model T chassis by the O. J. Childs Company is owned by Dr. Will Brooks, Founder & Past President Canadian Fallen Firefighters Foundation. PHOTO CREDIT: Dan Jasina Collection.
.

Shortly after Dan Jasina sent me the photo above, I received an email from Dr. Will Brooks with considerably more information about his 1913 Ford Model T Childs fire engine. Dr. Brooks emailed additional photographs and a copy of the story he submitted to Enjine! Enjine!.

Here in a condensed version is Dr. Will Brooks' story about his O. J. Childs fire engine.

"In April, 2003, Georges Potvin and I were driving back from Quebec City to Ottawa. We were winding along the Ottawa River through small villages and towns where fire engines of interest still may yet sleep in the old barn of a farmer. Georges noted that he had always had the dream to one day drive into a farmyard, meet the owner, and be shown an old and dusty barn in which stood an aging steam fire engine.

My reply to this musing was to say how much that would be wonderful if the engine were a motor-driven rig on which I could work without a boiler man's ticket. As with most fantasy, our dreams did not materialize, at least not then.

In August, Foundation business took me to Petawawa, and when finished with tasks there, I chanced to meet Richard Kramer who is part of the Petawawa Fire Department. As we spoke, it became clear to him I was the fellow who had purchased the old Bickle-Seagrave from his town in 1997.

Richard said he had something I might like to see. If I followed him, we could have a look right away. A few miles and twists and turns along country roads later, we arrived at an old barn. I went with Richard as he unlocked first one set of doors and then another until we were in a dusty space occupied by a large number of beautiful barn planks and a heap of heavy, black plastic covering.

Saying very little, Richard threw back one side of the cover to reveal a partially restored 1913 O.J. Childs Model T Chemical-Hose Wagon. He explained that his Dad had gotten it some 25 years or so before, had begun the restoration but had not been able to complete it before passing away. Out of respect, he and his brother Robert had just stored it.

Richard wondered if I might be interested in obtaining it. Interested, yes, able, well that meant some consultation. It was a great thrill to call Georges and tell him the story, that what we had just yarned about on the way from Quebec City had really happened. We took a trip to see the old Model T, imagine its final look and do some negotiating over price.

A few more weeks passed, and we struck a deal. Our friend Lionel hitched up his flatbed trailer, and we went off to bring the O.J. Childs to Gatineau where Georges could have first go at getting it running. Both his workshop and knowledge are much better than mine. Further, he had access to three friends who knew Model T engines from the days of their own youth; they were true fountains of knowledge as we worked on the finer points

The brass work on the Childs was plentiful. From rear step to radiator, brass seemed to have been the metal of choice. It was all badly in need of polishing, a task which took the better part of 6 months as piece by piece it was taken off, carted to Georges' workshop polished, and replaced. Finally, after many calls to Lang's Auto Parts in Massachusetts and deliveries of a starter and other small but essential items, the old rig was complete.

Unfortunately, we have been stymied, so far, in our search for the history of this rig. Although it was thought by the previous owners that it might have come from Florida, no solid evidence has proven that to be true. None of the usual archives have been able to help."

–Dr. Will Brooks

Combination Chemical and Hose Car

Mounted on Ford Pleasure Car Chassis

The O. J. Childs Company of Utica, New York began as a fire extinguisher company in 1896, building chemical fire apparatus in 1913. Childs advertised a Combination Chemical and Hose car mounted on a "Ford Standard Pleasure Car" equipped with a Hose Body. PHOTO CREDIT: Steve Hagy Collection

Ford Model T O. J. Childs Chassis. Matt Lee Collection

NOVEMBER 15, 1918 THE COMMERCIAL CAR JOURNAL

"Childs" Fire Equipment installed on any make of Truck Chassis!

MR. COMMERCIAL TRUCK DEALER:

You can save your city money, supply it with the very best quality Fire Fighting Apparatus and add to your own sales income by selling the "Childs" apparatus mounted on your chassis.

We mount the apparatus in our shops and return the chassis and equipment to you complete.

You make liberal profits.

"Childs" Fire Apparatus

offers the utmost in efficiency.

It has been adopted by the U. S. War Department for protecting army equipment in France, as well as munitions plants at home; by the New York City Fire Department, by Bethlehem Steel Company, The Ford Motor Company, and the largest railroads and manufacturers.

O. J. Childs Company, Utica, N.Y.

Childs Advertisement Commercial Car Journal November, 1918

1914 Ford Model T Fire Truck Chapel Hill (North Carolina) Fire Department

"Our T-Model was purchased in 1914 through the Ford dealership here in Chapel Hill. It was the town's first motorized fire truck, one of the first motorized fire trucks in the state, and one of seven built by Ford of that particular model.

I believe we still have the original invoice. The truck cost all of $600 when it was purchased nearly 94 years ago. It was basically a hose and equipment rig because it had no pump. It was restored in the late 1980s prior to my arrival here and we drove it in parades up into the mid 1990s.

The first 1914 Model T Touring bought in Chapel Hill, North Carolina served as the source for parts when the Chapel Hill Fire Department Model T was being restored. The "parts car" was later restored and is now on display in the Chapel Hill Museum about 20 feet from the fire truck.

We keep the Chapel Hill Fire Department Model T up on jack stands to keep weight off of the wooden spoke wheels." --Dan Jones, Fire Chief, Chapel Hill, NC PHOTO CREDIT: Dan Jones

1914 Ford Model T Equipment truck

"I have had the T since 1983. When I purchased it, it was not running good enough to drive. I had the engine overhauled by a local T mechanic. I kept breaking the wood spokes when I drove it so I bought 27 T hubs and wire wheels to improve the dependability.

I have been on Southeast Fire Department for 40 years this year, we started as a rural department and have become a suburban department due to most of the farm land being replaced with home developments (larger acreages) we do have one unincorporated town in our district, Walton, Nebraska. I wanted to restore it as a fire truck so I purchased or borrowed all the books I could find that had Ford T fire equipment in them so I could be as accurate as possible.

My wife was in the antique business for several years so we traveled the country looking for antiques to sell and fire truck equipment. The red light on the fender came from Wisconsin, the brass bell from Las Vegas, the fire lantern from Minden, Nebraska (home to Pioneer Village), old extinguisher on the fender from Council Bluffs Iowa and the rest of the equipment from various shops around the country. The old bunker coat, boots and helmet were mine when I was on the Lincoln Fire Department in 1969/1970.

The truck represents our fire department in parades, car shows, and annual open houses at the fire stations. I even made a fire call in it once when it was close by. It is a 1914 Ford model T." - Darrell Best. PHOTO CREDIT: Darrell Best

1914 Ford Model T - 1917 Smith Form-A-Truck

In 1917 the Washington Volunteer Fire Company, Washington, Missouri installed a Smith Form-A-Truck Conversion Kit on this 1914 Ford Model, along with the addition of firefighting equipment and 40-gallon soda acid system. This is believed to be the first motorized fire truck in a volunteer fire department in Missouri. The rig is restored, fully operational and retained by the Washington Volunteer Fire Company. PHOTO CREDIT: Dennis Maag

Chemical engines the most powerful fire fighters

American LaFrance Advertisement for Chemical fire engines.
PHOTO CREDIT: Steve Hagy Collection

April, 1916 SUCCESSFUL FARMING 99

How C.W. Printy saved a lot of money—

—a story every farm-owner should read.

"HERE was my situation. I needed a truck. I had many uses for one. I knew I could save both money and time by using one. But when I looked into the price of trucks, and found that the average one-ton truck cost about $1500, I could not see my way clear to buy one.

I also wanted a new Ford car for my family. My present Ford had done me good service and I knew it was still good for several years more.

"This is how my Ford looked after I had removed rear wheels, and part of body."

One time I thought it would be cheaper for me to put a new box body on my Ford car. But then I figured I would not have any car for my family use and would have only a make-shift delivery car. Also if I got rid of my Ford second-hand, it would not bring anywhere near its real value.

I could not buy both a new Ford and a new truck at the price I would have to pay.

About this time a friend of mine who owns a farm in Illinois bought a rear end construction especially designed to make a Ford car into a complete one-ton truck.

He had wonderful success with it. It carried loads far in excess of the guaranteed one-ton load.

It was just what I wanted. I wrote the Smith Form-a-Truck Company, Chicago, for full particulars. They immediately sent me catalog and complete literature showing the adaptability of the Smith Form-a-Truck for farm use. I showed these to my friends and all agreed the Smith Form-a-Truck was the very thing for all-around trucking and hauling which the average farmer has to do.

I could see by converting my Ford into a one-ton truck by the use of the Smith Form-a-Truck attachment, I could get full value out of my Ford. I would not have to lose anything on a "trade in."

I knew how powerful and dependable my Ford had been. Every one in our family could operate it without any trouble. I knew that most trucks used a lot of gasoline and oil, and were expensive in repairs. And I knew the Ford engine was most economical.

When I found I could get the Smith Form-a-Truck one-ton rear end construction for $350, I was astonished, knowing what it would do.

I noticed in the literature of the Smith Form-a-Truck Company that some of the biggest concerns in the country were using them, such as the Standard Oil Company, Armour, Morris & Company, Indian Refining Company, big department stores and cities, including the city of Chicago.

I knew that these big concerns had figured haulage and delivery costs down to a fine point. They knew these things better than I did.

So I sent and bought one of these Smith Form-a-Truck attachments. And you may be sure I could hardly wait for it to arrive.

It finally came. It was a fine piece of work. The steel frame fitted right over the frame of my Ford car up to the front end. It bolted together securely, thus doubling the strength of the entire frame.

The frame work was a 4-in. channel steel, reinforced with cross members and corner braces. It was 14 feet long and accommodated a standard 9-foot loading body.

In three hours, with the help of another man, I converted my Ford car into a one-ton truck. None of the famous Ford mechanism was disturbed. The rear axle of my Ford became the jackshaft of my truck.

There were specially designed truck springs, a real truck axle 2x1⅝ inches in size, and all the strength of a big truck.

The tires were the famous Firestone solid rubber, pressed on. To make the drive they used chains of a heavy roller type.

I never saw such a surprised lot of men as my neighbors were when they realized that at the small expense of $350 I had turned my good old Ford into the finest one-ton truck I could possibly get.

And one of the best points about the whole proposition is that now I can afford to buy a new Ford for my family.

I can't tell you how thoroughly satisfied I have been in every way with this Smith Form-a-Truck proposition.

I have hauled 2300 pound loads of corn from my farm to Covington over soft gravel roads, and yet the Smith Form-a-Truck has stood up like a major. It used to take 2 wagons and 4 horses to do the work that I am now doing with my Smith Form-a-Truck. If you knew the condition of the roads around here you would realize more and more just what it is doing for me.

"This is the Smith Form-a-Truck rear end construction, complete, just as it came to me."

No one had to tell me anything about the advantages a truck would be to me. I knew that, but I also knew that I could not afford to pay the price that the average one-ton truck cost.

And then there was the other problem of what to do with my Ford car. You helped me solve the whole proposition, and at a price way, way, way under what all my friends told me I would have to pay in order to get any kind of a truck at all. Now they are as much surprised as I am. Any number of them are figuring on turning their present Ford cars into Smith Form-a-Trucks, just as I have done, and buy new Fords this Spring for their pleasure riding.

If anybody should ask you, you can just tell them that C. W. Printy of Covington has got one of the finest trucks in the whole State of Indiana. And you helped me to do it."

"——and in three hours time I had a truck that cost half what other one-ton trucks cost, and did what it formerly took four horses and two wagons to do."

C. W. Printy

$2,000,000 sold in last 30 days—mostly on Farms

This is the best answer to what a Smith Form-a-Truck has DONE and will DO. Farmers from coast to coast, all of them most exacting buyers, are buying Smith Form-a-Trucks for all kinds of work on farms and ranches. They buy because the Smith Form-a-Truck offers wonderful value. We never could have built our tremendous business, in so short a time, if we did not give the best for the money.

$350 and a FORD makes this guaranteed 1-ton Truck

Put It Up to Us

Self interest demands that you investigate.

Tell us your hauling problems, and our transportation engineering department will give you valuable information and common sense advice, without any obligation on your part. If your hauling is costing you too much, you want to know it. You then need the Smith Form-a Truck.

Put it up to us, write or wire us today. Get our illustrated descriptive folder, "It solves your delivery problem," containing full detailed specifications and other information you want to know. Address

Smith Form-a-Truck Co.
919 Adams Express Bldg.
Chicago, Ill.

IMMEDIATE DELIVERIES
No doubt about it. We will make 50,000 jobs for 1916. Our big plant is now working feverishly night and day. There will be no slow-down. No let-up in our present quick service to buyers.

Smith Form-a-Truck

Smith Form-a-Truck Co.
919 Adams Express Bldg., 115 S. Dearborn St., Chicago

Kindly send me your folder, "It Solves Your Delivery Problem," and put before me facts and figures proving that Smith Form-a-Truck gives the cheapest delivery cost in the world.

Name..

Street Address..

City.............................. State...............

***April 1916 Successful Farming* Magazine Smith Form-A-Truck Advertisement**

In 1915 the Smith Form-A-Truck Company offered a "rear end construction complete" for $350 to modify the Ford Model T into a one ton truck. Like local farmers, many fire departments converted the Model T car into trucks for use as fire apparatus. PHOTO CREDIT: John Purkey III

1914 Ford Model T Chemical

The Hempstead Fire Department, Long Island, New York has owned this 1914 Ford Model T since it was purchased new in 1914. Shown here at a muster in 2007, the rig was restored in 1996. PHOTO CREDIT: Les Arsell.

Hempstead Model T History

"We are glad to see that somebody is taking interest in our T the way we do. Here's the story to the best of my knowledge. As I said before this is an original piece of apparatus of the Hempstead Fire Department, it stayed in service until the late 1920's.

Between the 1920's and the 1950's it sat in an old barn at the Village of Hempstead water plant along with an 1882 Silsby Steamer to keep it company. (By the way we still have that too.) In 1957 the Hempstead FD had its 125 Anniversary and the T was brought out and was part of the celebration even though it was not in very good shape. The barn where it was kept had a hole in the roof and weather had done a number on the wood and paint. Even though, it was still running and was used in the parade and then used for display during fire prevention week for quite a few years and then was stored in fire headquarters after that.

Hempstead Fire Department's 1914 Ford Model T Chemical shown here at the Garden City Train Station in 1914. PHOTO CREDIT: Les Arsell

Around 1974 one of the chiefs at the time, Jim Callahan, along with my father who is a member of Engine 2 out of headquarters wanted to fix the T up. They put new plugs, tires, paint, gold leaf and just some TLC to it. After that one of our old timers an Ex-Chief named Dan Redfield also from Engine 2 who was now the "T's keeper began taking it to parades. It was in that condition and remained that way until about 1995 when a group of us decided to refurbish the old girl.

Myself, Dan Redfield, Les Arsell Sr. Lael Von Elm, Bill Sielski, and quite of few others from the company and with the help of a friend who has a T of his own and a lot of knowledge, George Balduf, dismantled the car and redid just about everything.

We rebuilt the motor which needed new pistons, had the coils fixed, remade by hand the fire wall which under all the coats of paint we found this beautiful light colored wood which we think was oak, with a reddish varnish on it which was remade to look like the original and remade the floor boards and seat. A new radiator was installed along with a water pump. The transmission was in bad shape and we rebuilt it with parts from another old transmission and installed new bands in it.

We sent the wheels to California to be redone. The rear is a special rear so the wheels are a different size than your normal T. While they were in California there was a huge mudslide which wiped out the shop where the wheels were. We were at a loss for words because we could not locate the same size wheels. Before we sent them out we had the sense to mark them. Well, God must like us because where the slide took everything they found a bunch of wheels but had no idea who the belonged too. We explained about the markings we had put on them and believe it or not we got them back!! We did all the work ourselves except for the wheels.

We have won so many trophies and awards with this car and we are very proud of it. Also she is not a trailer queen. We drive it to most of the shows on the Island. We have just replaced the tires again with white ones for the looks, originally had grey. A lot of people tell us that the head lights have been replaced because they are not carbon tech and that's what a 14 had. When the truck was new, which is when the old pictures were taken then were electric. If you look at the pictures the carbon tech canisters are there but no headlights were ever found.

The tank on the back was originally nickel, but was painted at an unknown time. We were going to try to bring it back to nickel then decided it looked just as well painted. We were also going to static test it to see if we could get it to work but we decided against it for safety reasons.

When my father was Chief in 1984 he bought the T from the village for a dollar. Now if that's not a deal, what is? We had it appraised right after finishing it in 1996 and because it's a specialty car and the condition it was in, we were quoted around 45,000.00 dollars. Not bad for a bunch of firemen."

Lester J. Arsell, Ex-Chief-Engine 2, Hempstead Fire Dept

The American-La France Combination Chemical and Hose Car

American LaFrance Advertisement for Chemical fire engines.

PHOTO CREDIT: Steve Hagy Collection

1909/1915 Ford Model T Bain

The Bain Wagon Company, Woodstock, Ontario, Canada converted this 1909 Model T Ford into a plant fire truck in 1915. PHOTO CREDIT: Walt McCall Collection.

1914 Ford Model T American La France Hose Car

Lynnfield, MA Fire Department's 1914 Ford Model T American LaFrance Chemical

The Lynnfield, MA Fire Department's 1914 Ford/American LaFrance Model T Chemical wagon was donated to the town by the proprietor of the local house of questionable repute. As you could imagine, quite a debate raged about its acceptance until cooler heads prevailed. This photograph is in front of the Lynnfield Chemical House # 2, and was taken when the T was fairly new as the configuration of this house was changed in 1919-1920. The truck today is almost identical in appearance to this photo. Chemical #2 was used up into the 1950's and is still maintained for parade duty. Chemical #2 has American LaFrance serial number F106. CREDIT: Paul Romano, Chief Lynnfield, MA FD

AMERICAN-LAFRANCE CHEMICAL
FOR SMALL TOWNS AND VILLAGES

ON STANDARD FORD CHASSIS

1914 Ford Model T American LaFrance Chemical Car

The Lynnfield, MA Fire Department still owns this 1914 Ford/American LaFrance Model T Chemical wagon. PHOTO CREDIT: Walt McCall Collection.

1915 Ford Model T Chemical

PHOTO CREDIT: Collection of Thomas L. Herman

1915 Ford Model T Double Tank Chemical

The fire department is not identified, although the engine is lettered M.F.D., Worcester, Pennsylvania. PHOTO CREDIT: Tod Wirth Collection

Circa 1915 Ford Model T Chiefs Car

Southampton Fire Chief's Runabout PHOTO CREDIT: Collection of Thomas L. Herman

1915 Ford Model T American LaFrance Chemical

This single tank chemical is believed to be the first one built b American LaFrance on the Ford Model T chassis. PHOTO CREDIT: Walt McCall Collection

1915 Ford Model T Gas and Foamite Supply No. 1

In approximately 1915, the City of Richmond, VA acquired this gas and Foamite supply truck, which was built on a Ford Model T chassis. This photograph was taken in front of Station #12, located at 229 West Grace Street. PHOTO CREDIT: Collection of Thomas L. Herman

1915 Ford Model T American LaFrance Chemical F107

"This unit was delivered to the Las Cruces NM fire department in the early fall of 1915, and remained in service until around 1934 because it still had the exempt NM 1934 license on it when I purchased it in 1999 from Joe Melendrez, who was a volunteer fireman in Las Cruces, who purchased it from the City in 1937 after it sat in their scrap yard for several years. Melendrez kept it in his barn and occasionally drove it while his children were small, then it just sat until I purchased it from him in 1999." – Wes Melo PHOTO CREDIT: Wes Melo

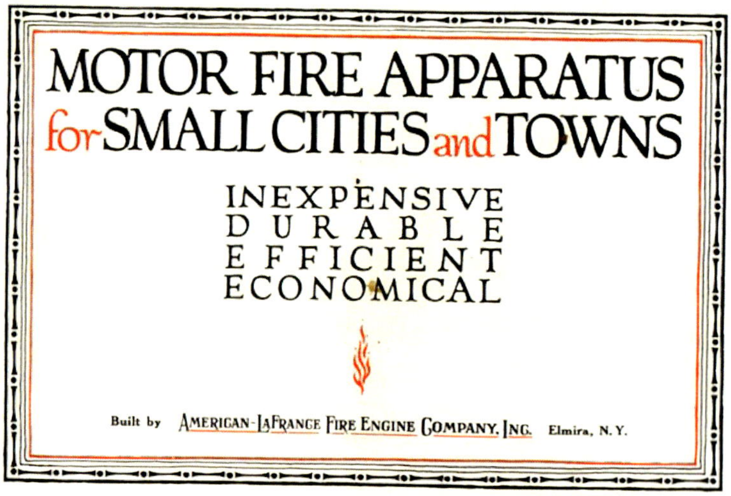

American LaFrance Advertisement

PHOTO CREDIT: Steve Hagy Collection

1915 Ford Model T with American LaFrance Equipment

This Model T is on display in the Chicago offices of Fireman's Fund Insurance Company. PHOTO CREDIT: Fireman's Fund Insurance Company

County of Los Angeles Fire Museum Association's 1915 Ford Model T

This chemical/hose car is typical of vehicles purchased by small communities early in the 20th Century. These light-weight rigs usually had no pump carried a quantity of hose that could be used to provide a supply line between a hydrant and a fire pumper. PHOTO CREDIT: County of Los Angeles Fire Museum Collection

1915 Ford Model T American LaFrance Double Tank Chemical

Central Ohio Fire Museum's 1915 American LaFrance Double Tank Chemical was restored by local Model T and Model A Clubs in 2005-2006. PHOTO CREDIT: Tom Herman Collection

1916 Ford Model T Chemical

This 1916 fire engine is owned and displayed by the Schuylkill Historical Fire Society museum in Shenandoah, PA. PHOTO CREDIT: Michael Kitsock.

1916 Ford Model T American LaFrance Double Tank Chemical

This 1916 American LaFrance Ford Model T double tank chemical, Registration F-133, was delivered to the Niagara Fire Company, Merchantville, New Jersey and was featured in "Owner's Pride" in Enjine! Enjine! 1985 Volume 1. PHOTO CREDIT: SPAAMFAA ENJINE! ENJINE!

1916 Ford Model T Double Tank Chemical Engine

This brass-era Model T beauty, with its painted wooden spoke wheels and white tires is owned by Joseph Findysz. The rig was modified into a chemical truck in March 1916, by James Boyd and Brothers, Incorporated. The Boyd's were known for constructing small trucks, hose carts, and chemical carts in the early twentieth century. This particular chemical truck was used in the railroad yards of California. PHOTO CREDIT: Bill Killen Collection

1916 Ford Model T Double Tank Chemical,

Wyoming Hook and Ladder Fire Company's Double Tank Chemical car. PHOTO CREDIT: Collection of Thomas L. Herman

1916 Ford Model T American LaFrance Double Tank Chemical

Campbell Hall (New York) Fire Department's Engine No. 1 photographed at a June 25, 1982 muster. PHOTO CREDIT: Tom Herman Collection

Yuma, Arizona's 1916 Ford Model T American LaFrance

Yuma, Arizona's first self-propelled fire apparatus was this 1916 American LaFrance on a Ford Model T with a Smith "Form-a-Body truck conversion kit. This historic piece of Yuma Fire Department equipment is still in service as a ceremonial piece. PHOTO CREDIT: Vintage Fire Brigade, Tucson, Arizona.

SPAAMFAA "T" Party

What is 'SPAAMFAA"? The Society for the Preservation & Appreciation of Antique Motorized Fire Apparatus in America has more than three thousand members in over fifty chapters, with membership world wide. These five Ford Model T fire engines lined up at the 2004 SPAAMFAA National Muster held in Frankenmuth, Michigan. PHOTO CREDIT: Walt McCall

Ford Model T American LaFrance Single Tank Chemical

American-LaFrance touted their Single Tank Chemical Car as "the ideal car for the fire chief of a small city. It gets him to the fire at the earliest possible moment and if he arrives before the regular apparatus it often enables him to nip the flames in the bud. It therefore makes the chief doubly efficient." American-LaFrance advertisement – PHOTO CREDIT: Steve Hagy Collection

Ford Model T American LaFrance Double Tank Chemical

American LaFrance factory photograph circa 1916 of a double tank chemical system. PHOTO CREDIT: Steve Hagy Collection.

"BUFFALO" FORD CHEMICAL CAR

SPECIFICATIONS

CHASSIS

Ford Standard One Ton.

CHEMICAL TANKS

3 BUFFALO Standard 35 gal. steel tanks. The three tanks so placed as to form two sets. Set No. 1 consists of two tanks side by side, but laterally on body over rear axle, so connected that one continuous stream of chemical can be maintained. Third tank located in front of set No. 1, an individual unit.

3 Acid Canisters.
3 Soda Canisters.

CHEMICAL HOSE

300 ft. ¾" Chemical hose, coupled in 50 ft. lengths.

HOSE CARRIERS

2 Perforated steel baskets equipped with rollers or reels.

NOZZLES

2 Eccentric brass shut-off nozzles.

LADDERS

1—12 ft. roof ladder.
1—24 ft. extension rapid-hoist ladder.

EXTINGUISHERS

2—3 gal. tip-over type.

FIRE EQUIPMENT

2 Dietz Fire Dept. Lanterns.
1 Fire Dept. Axe.
1 Steel Crowbar.
1 10' Pike Pole.
2 Hardwood nozzle plugs attached to rear step.

ALARM SYSTEM

1 Electric or hand siren.

PAINTING AND LETTERING—Painting done in first class manner with the best wearing colors and varnish. Lettering and decorating in gold leaf.

Buffalo Fire Appliance Corporation Advertisement

Buffalo built pumpers, hose cars and chemicals on the Ford Model T chassis. This ad features one of the two chemical engines built for Eden, (Erie County) New York. The Eden Volunteer Fire Department, organized in July 1923, consisted of two departments and two pieces of chemical apparatus. PHOTO CREDIT: Steve Hagy Collection.

"BUFFALO" FORD PUMP, CHEMICAL, HOSE CAR

Triple Combination

SPECIFICATIONS

CHASSIS
Ford Standard One Ton.

PUMP
Rotary mounted under the driver's seat.

HOSE BODY
Capacity 800 feet.

CHEMICAL TANK
1 BUFFALO Standard 35 gal. steel tank.
1 Acid Canister.
1 Soda Canister.

CHEMICAL HOSE
150 ft. ¾" Chemical hose coupled in 50 ft. lengths.

HOSE CARRIER
Perforated steel basket equipped with rollers.

CHEMICAL NOZZLE
1 shut-off eccentric non-corrosive.

ALARM SYSTEM
1 Electric or Hand Siren.

LADDERS
1—12 ft. roof ladder.
1—24 ft. extension rapid-hoist ladder.

SUCTION HOSE
2—10 ft. lengths 2½" smooth-bore.

STRAINER
1—2½" Strainer of solid cast brass, polished.

EXTINGUISHERS
1—2½ gal. tip-over type.

FIRE EQUIPMENT
2 Dietz Fire Dept. Lanterns.
1 Fire Dept. Axe.
1 Steel Crowbar.
1 8' Pike Pole.
2 Hardwood Nozzle plugs attached to rear step.

PAINTING AND LETTERING—Painting done in first class manner with the best wearing colors and varnish. Lettering and decorating in gold leaf.

Buffalo Fire Appliance Corporation Advertisement

This Buffalo Triple Combination was built for the Riverdale Fire Department (state unknown) on the Ford Model T chassis. PHOTO CREDIT: Steve Hagy Collection

"BUFFALO" FORD CHEMICAL CAR

SPECIFICATIONS

Chassis
Ford Standard One Ton.

Chemical Tanks
2 BUFFALO Standard 35 gal. steel tanks.
2 Acid Canisters.
2 Soda Canisters.

Chemical Hose
150 ft. ¾" Chemical hose coupled in 50 ft. lengths.

Hose Carrier
Perforated steel basket equipped with roller.

Nozzle
1 Eccentric brass shut-off nozzle.

Ladders
1—12 ft. roof ladder.
1—24 ft. extension rapid-hoist ladder.

Extinguishers
2—3 gal. tip-over type.

Fire Equipment
2 Dietz Fire Dept. Lanterns.
1 Fire Dept. Axe.
1 Steel Crowbar.
1 10' Pike Pole.
2 Hardwood Nozzle plugs attached to rear step.

Alarm System
1 Electric or Hand Siren.

Painting and Lettering—Painting done in first class manner with the best wearing colors and varnish. Lettering and decorating in gold leaf.

Manufactured by

BUFFALO FIRE APPLIANCE CORP.

Office and Factory
44 Central Avenue, BUFFALO, N. Y.

Western Sales Office
417 So. Dearborn Street, CHICAGO, ILL.

Buffalo Fire Appliance Corporation Advertisement

This ad features a "Buffalo Ford Chemical Car" for what appears to be the Town of Line, believed to be in Erie County, New York. PHOTO CREDIT: Steve Hagy Collection

Turlock, California's 1917 Ford Model T Chemical

PHOTO CREDIT: Brian White, Turlock City Fire and Emergency Services

"This 1917 Model T Ford Chemical was purchased for the Turlock Fire Department in 1918 and remained in service as a front line engine until around 1922 when it was replaced by a 1922 American LaFrance pumper.

The Model T was used in parades and various events until 1963 and was in use by the City of Turlock for various public events throughout the 1960's. In the early 1970's it was being used at the city corporation yard to pump at the city sewer farm.

A process to acquire and restore the Model T to its firefighting days was started by then Fire Chief C. Chittock in 1978. Apparently, the Fire Department members were each given parts to clean, restore and rebuild. It took about a year to restore it and a second restoration, re-painting and gold leaf work was completed around 1985 to its current appearance.

The pump is a "Bean" and only a guess but it does have two 2 1/2 outlets so it may have a 250 GPM capacity. The pump is chain driven. The Model T is functional to drive and the Turlock Fire Department uses it in parades and other public events.

Although Turlock Fire Department does not pump with it the unit, it is complete and could function if desired. Additionally, in the 1980's it competed in several California Fire Musters and won top place trophies two years in a row at a Fire Muster held in Redwood City, California when Marine World (Theme Park) was at that location but the trophies were perpetual and are not in our possession."

Captain Brian White, Turlock City Fire and Emergency Services

1917 Ford Model T Fire Chief Car w/ American LaFrance Chemical Unit

This 1917 Ford Model T Fire Chief Car is owned by James F. Bacon, Golden Era Automobile Association, Orting, Washington

1917 Ford Model T American LaFrance Chemical Unit

This unit was powered with a four-cylinder, in-line, water-cooled 20 hp engine with a displacement of 176.7 cubic inches. The body and equipment were furnished and installed by American-La France. This vehicle was used by the Farmington, Maine, Fire Department and is completely original and has never been restored. PHOTO CREDIT: Owls Head Transportation Museum

This 1917 Ford Model T Howe was one of 168 the US Army purchased for assignment at Army camps throughout the United States and Europe during WWI. PHOTO CREDIT: US Army.

1917 Ford Model T Howe Camp Upton, NY

1917 Ford Model T Howe Camp Custer, Michigan

These Ford Howe pumpers were assigned to the Base Hospital at Camp Custer, Michigan Camp Custer was where the 85th Division was trained during World War I. Troops from Michigan and Wisconsin made up this Division. PHOTO CREDIT: Bill Killen Collection.

1917 Ford Model T Howe Fort Sill, Oklahoma

This is an early 1917 engine Howe built for the Army Quartermaster Corps, since the words "Pat. Applied For" are beneath the Howe name on the tank. PHOTO CREDIT: Matt Lee Collection

Harbor City Fire Department's Ford Model T Pumper

It is unusual to find photographs of Model T fire apparatus on the fire ground. I am unable to determine if there is a fire, the engine is hooked to the hydrant, the lines are charged and the driver is operating the pump from the driver's seat. PHOTO CREDIT: Matt Lee Collection

Circa 1916-18 Ford Model TT Chemical Hose Wagon

The manufacturer of the Salem Auto Museum and American Shop's Model TT is unknown
PHOTO CREDIT: Tom Herman Collection

Henderson Fire Dept.
1921
Calvin Scott,
Fred Wilson,
Leslie Cross,
Kensel Sprague,
Herman DeLaVergne,
Gaylord Montague,
Lewis Woodruff,
Birt Slack,
Howard Tifft,
Letcher Allen,
Lee Whitney,
Bernard Sprague,
LaMort Burnham,
Leo Sprague,
Docky Howard,
Herbert Hunter

Henderson, New York's 1921 Ford Model TT Waterous Pumper

The engine of this Ford Waterous pumper was cooled by water piped from the tank to the engine when operating the fire pump. PHOTO CREDIT: Henderson Historical Society

1917 Ford Model T Chemical Unit Grant Park Fire Protection District

PHOTO CREDIT: Bill Friedrich

1917 Ford Model T Hale 250 GPM Pumper

Malden Miller of Hampstead, Maryland owns this 1917 Hale 250 GPM Rotary Gear pumper, Hale #345, the oldest known Hale pump in existence. This rig was in service in South Hampton, PA until 1937 and pumps at musters on a regular basis. PHOTO CREDIT: Malden Miller

Midway, Kentucky's 1917 Ford Model T Howe

This green on red color combination is reminiscent of the colorful designs and painting on hand and horse drawn pumpers and hose carriages. PHOTO CREDIT: Jerry Sudduth

Fire Equipment Limited Postcard Advertisement

The fine print at the top of the card reads "One of our most up-to-date one ton fire trucks built". Fire Equipment Limited, Montreal, Canada, manufacturod several fire engines on the Model T chassis. PHOTO CREDIT: Steve Hagy Collection

Livingston Manor, New York's Ford Model T Howe

Livingston Manor Volunteer Fire Department's Engine 001. PHOTO CREDIT: Jim Atkinson

1917 Ford Model T Soda Acid

St Tammany Parish (LA) Fire District #4. PHOTO CREDIT: Tom Herman Collection

Howe-Ford Model No. 3 Double Tank Chemical on Ford Model T Touring Chassis

This Howe Fire Apparatus Company advertisement features the Bowdoin Fire Department's double tank chemical. PHOTO CREDIT: Steve Hagy Collection

1917 Ford Model T Prospect Deluge Model C with hose reel

This Ford Prospect Deluge Model C was built for Fairport, Ohio shown here in front of the Prospect Fire Department engine house. PHOTO CREDIT: Tod Wirth Collection

1917 Ford Model T Howe pumpers at Fort Benjamin Harrison, Indianapolis, Indiana

Howe built 168 of these pumpers during 1917 and 1918 for the US Army Quartermaster Corps. Note the ugly dog on the left. Many of the Army's Howe pumpers saw service in municipal fire departments. PHOTO CREDIT: Steve Hagy Collection

Violetville Volunteer Fire Department

Violetville Volunteer Fire Company founded in 1906, 6th oldest volunteer fire company in Baltimore County, Maryland, shared in a dark day for Baltimore City firefighters. In the early twentieth century the St Mary's Industrial School for Boys was destroyed by a devastating inferno in which many Baltimore City firefighters were injured and two made the ultimate sacrifice. Violetville's 1917 Ford Model T, which sat in disrepair from 1955 until 2000, was in service at the school fire and operated at the scene of a fire that ruined the childhood home of the legendary baseball hero Babe Ruth.

In 1917 the Violetville Volunteer Fire Department purchased a Ford Model T chassis with no body and purchased a Smith Form-a-Truck device, Serial No. 26404 and added a chain-driven transmission.

The body was constructed by a local business and accessories and rails were produced by local companies such as F. L. Anderson, a Baltimore company serving the fire service since 1919. In the beginning the engine had no pump and was used only as a hose wagon. In the mid 1920's an 1886 Holloway Soda Acid Chemical Tank was installed above the driver's seat. The addition of this on board extinguishing system and a 100 foot chemical "attack" hose improved firefighting capability. The lighting package for Engine #1 was six kerosene lanterns that were lit each evening by volunteers.

Violetville Volunteer Fire Department members circa 1918

The 1917 Model T remained in service until 1930 when a Pierce Arrow Chemical engine was purchased. However, the T continued to make parades and festivals in various communities throughout Maryland, including a 15 hour one way trip to the Maryland State Fireman's Association convention in Frostburg, in Western Maryland.

In 1955 after a fire destroyed the Violetville Firehouse, the Model T was moved to a member's shed, where it sat for five years. In the early 1960's it was stored outside the firehouse under canvas tarps, metal plates and sheets of tin. Finally, in 1969 it was moved to a shed added on to the firehouse where it was stored until 2001.

Violetville's 1917 Ford Model T, driven by John Purkey, Jr., leads the Maryland State Fireman's Association parade in June 1947 at College Park, Maryland

After the deaths of Chief John Purkey, Sr and his brother Emory Purkey in the early 1980's, the company honorably removed the chief's position and Captain John Purkey, Jr. became the department's chief officer. Captain Purkey helped undertake the restoration of the Model T in 2002 and retired as Violetville Volunteer Fire Department's Captain in 2006 after more than forty years of service.

1917 Ford Model T Smith Form-A-Truck

Captain John Purkey, Jr. driving Violetville Volunteer Fire Department's Model T in the Maryland State Fireman's Convention parade in Ocean City, Maryland, June, 2006. Fifty years earlier Captain John drove this truck in the 1947 convention parade in College Park, Maryland.

Violetville VFD's 1917 Ford Model T carried John Purky to the cemetery

Captain John Purky Jr rode it, drove it, restored it, and was buried off it and his family had the 'T' engraved into his headstone. VIOLETVILLE PHOTO CREDITS: John Purky III

1917 Ford Model T American LaFrance F-233

This American LaFrance Type 32 Double Tank Chemical was originally sold to the Phillips Fire Company, Weirton, West Virginia. PHOTO CREDIT: Tod Wirth Collection

1917 Ford Model T American LaFrance F-218

"This American LaFrance Type 32 was red originally but was restored by a retired Denver firefighter (their equipment is white). It is believed it came from New Hampshire." – Tod Wirth PHOTO CREDIT: Tod Wirth Collection

1917 Ford Model T Hose & Chemical Wagon

The Toledo, Ohio fire department placed this Hose & Chemical Wagon in service in 1917 as a "purpose built spare". PHOTO CREDIT: Toledo Firefighters Museum.

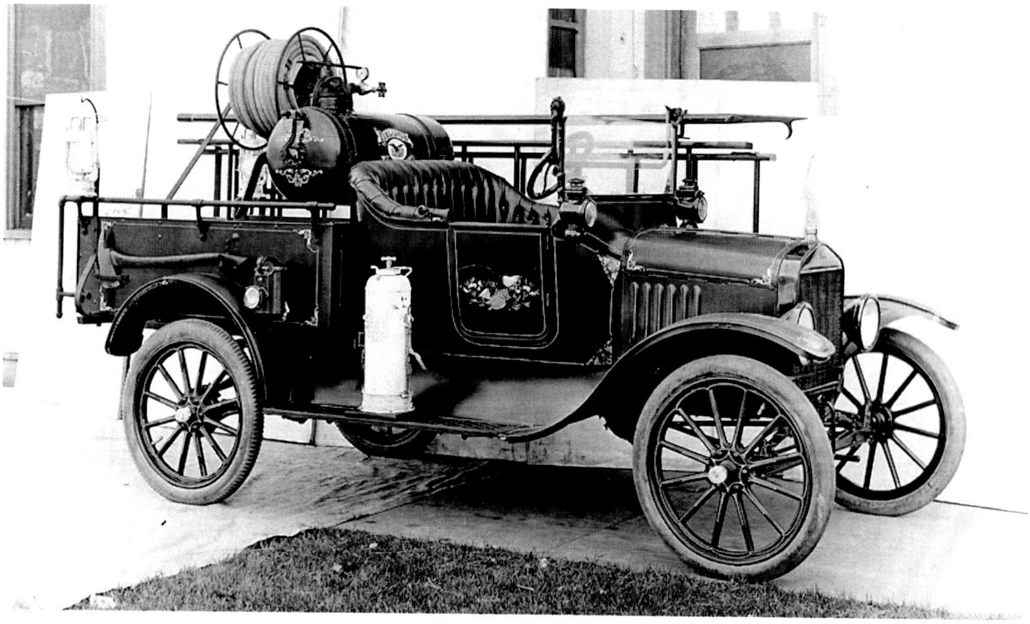

1917 or 1918 Ford Model T Prospect Model CH Hose & Chemical Wagon

The Toledo, Ohio fire department placed this Hose & Chemical Wagon in service in 1917 as a "purpose built spare". PHOTO CREDIT: Toledo Firefighters Museum.

1918 Ford Model T Hose & Chemical Wagon

The builder and year of manufacture of Mountain Lakes (New Jersey) fire department's hose and chemical is unknown. From examination of the features it appears to be a 1918 Model TT. PHOTO CREDIT: Jerry Suddath Collection

1918 Ford Model T Pirsch Hose Wagon

Chicago Fire department's Engine 126 appears to have stretched out a 2 1/2 inch hose line. The spare tire is mounted behind the driver's seat and a gong placed beneath the driver's seat. PHOTO CREDIT: CAPCO photo from the Steve Hagy Collection.

1918 Ford Model T Pirsch Hose Wagon

Chicago Fire department's Engine 82. There are slight differences between Chicago's Engine 82 and 126. This rig has the locomotive type bell and a hose basket mounted above the hose bed. PHOTO CREDIT: CAPCO photo from the Steve Hagy Collection.

1918 Ford Model T Prospect Model C

C. H. Sutphen & Son were factory representatives for the Prospect Fire Engine Company and these photographs are left and right side views of the Ford Model T Prospect Model C Chemical engine mounted on a passenger car chassis. PHOTO CREDIT: Tod Wirth Collection

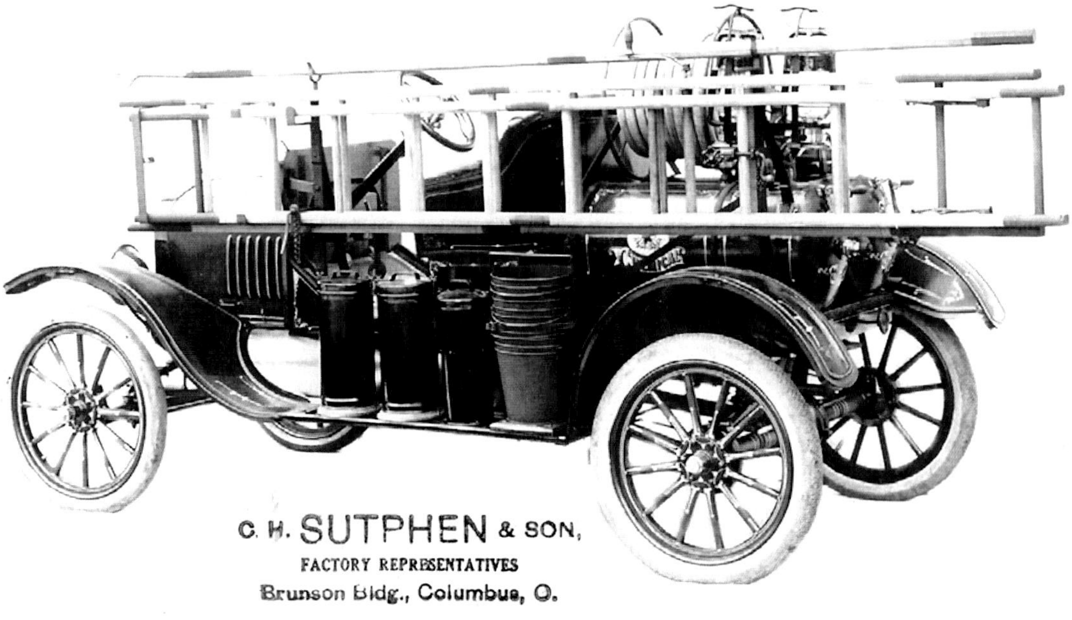

1918 Ford Model T American LaFrance Chemical Engine

The Falls Church Volunteer Fire Department has owned and operated American LaFrance Double Tank Chemical #F-393 for more than ninety years. Photo below shows the valves, gauges and discharge connection. PHOTO CREDITS: Robert Rohrback

1918 Ford Model T Howe 350 GPM Triple Piston Pump

This engine is one of hundreds of fire apparatus exhibits in the Hall of Flame Fire Museum in Phoenix, Arizona. Originally built for the U.S. Army in 1918 it was sold to the Germantown, Illinois fire department in 1920. PHOTO CREDIT: Bill Killen Collection

Ford Model T Chemical

Citizens Hose Company #1, Dillsburg, Pa. PHOTO CREDIT: Tom Herman Collection

Ford Model T Hose Wagon

Thornwood Fire District, organized 1899. PHOTO CREDIT: Tom Herman Collection

WATEROUS FIRE ENGINE COMPANY
Fire Fighting Apparatus
ST. PAUL, MINNESOTA

Ford Model T Waterous Pump

Factory photo of circa 1918 Ford Model T chassis with Waterous pump mounted on the frame beneath where the driver's seat would be located. PHOTO CREDIT: Steve Hagy Collection

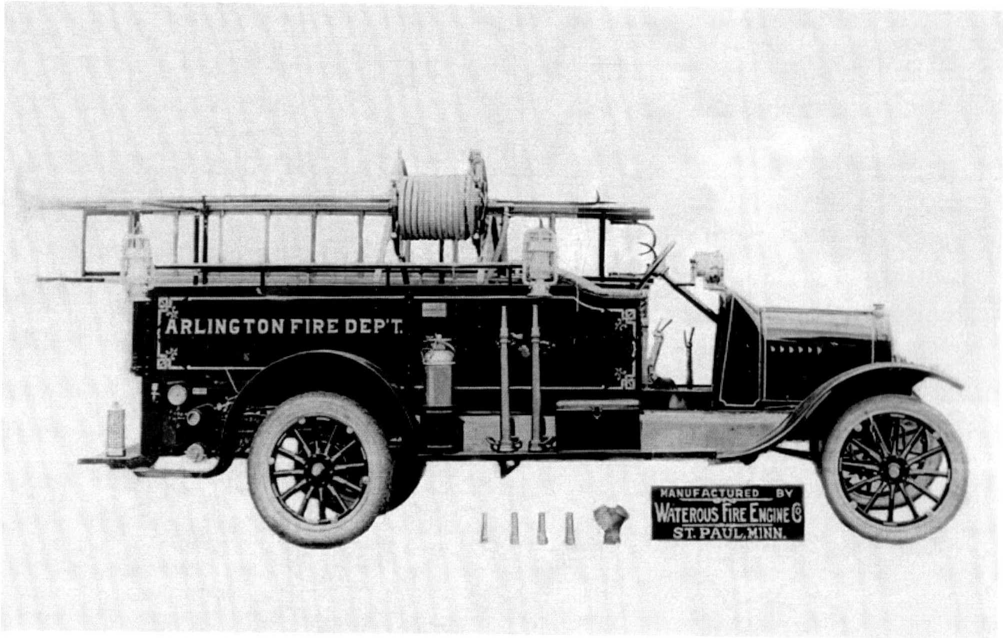

Ford Model T Waterous Triple Combination Firefighter

Circa 1918 Factory photo of a Ford Model T Waterous triple combination. Note the pump under the rear of the hose body and the reel mounted above the hose body. The state this "Arlington Fire Department" pumper was built for is unknown. PHOTO CREDIT: Steve Hagy Collection

1918 Ford Model T Thibault Combination Hose & Chemical

The first motorized fire apparatus built in Sorel, Quebec, Canada, by the famed Canadian Company was for the Campbellton Fire Department in New Brunswick, Canada. PHOTO CREDIT: Walt McCall Collection

1918 Ford Model TT

U.S. Naval Proving Ground, Dahlgren, VA.'s first fire truck was built on this Model TT chassis. The engine was equipped with a chemical tank and booster hose, Dietz lanterns, Sterling hand crank siren and two sections of hard suction hose. CREDIT: U.S. Navy Photo

1919 Ford Model T Howe Piston Pumper

Ca. 1919 Ford Model T Howe Piston Pumper served the Fox Lake, Wisconsin community. PHOTO CREDIT: Matt Lee

1919 Ford Model T Howe Piston Pumper

Ca. 1919 Ford Model T Howe Piston Pumper with Hose Trailer. PHOTO CREDIT: Matt Lee

1919 American La France Ford Model T

This 1919 American La France Type 75 Triple on a Ford Model T Chassis, registration number 2766, was shipped to the Norfolk Navy Yard, Norfolk, VA on September 30, 1919. PHOTO CREDIT: Bill Killen Collection

Advertisement for American LaFrance Chemical on Standard Ford Chassis

PHOTO CREDIT: Steve Hagy Collection

Ford Model T Howe Pumpers Ready for Delivery

These nine Ford Model T Howe pumpers are part of the 168 units built for the Army Quartermaster Corps 1917-1918. PHOTO CREDIT: Matt Lee Collection

1919 Ford Model T Front Mount Pump

This 1919 Model T Ford (manufacturer unknown) was driven in the London to Brighton Commercial Vehicle Road Run May 2002. PHOTO CREDIT: Bill Killen Collection

1919 Ford Model T American LaFrance

Goodwill Fire Company, Jacobus, PA owns this 1919 Ford Model T American LaFrance with a Gould hand pump. PHOTO CREDIT: Howard Miele III

1919 Model T Front Mount Pump

This truck marked Wayne County Fire Department is on display at the Westland Museum in Michigan. PHOTO CREDIT: Bob Rebtoy

1919 Ford Model TT Prospect Model CH

This un-restored Ford Prospect Chemical & Hose served New Philadelphia, Ohio. The "Deluge Chemical" decals are still legible. PHOTO CREDIT: Tod Wirth Collection

1919 Ford Model T District Fire Chief's car

District Chief Keller and Clarence "Taffy" Ihram in one of two Ford Model T Runabouts the Toledo Fire Department provided for newly created District Chiefs. The Model T's were off the Toledo Fire Department roster by 1926. PHOTO CREDIT: Toledo Firefighters Museum.

Ca 1915 Ford Model T Chemical Wagon

Believed to be Santa Clara, California. PHOTO CREDIT: Bill Killen Collection

1919 Ford Model T

This truck was purchased in 2006 from Thayer, KS Fire Chief Larry Myers.

In the early 1900's, the federal government initiated a program that gave fire trucks to small communities. Thayer received one of these, also known as "soda trucks" because a certain chemical was added to the water to create pressure from the water hose.

They used it for many years before it was sold to a local resident who drove it in parades. Eventually it fell into disrepair and was purchased by Fire Chief Larry Meyers who spent many years restoring it to its original condition.

Fireman's Fund found the truck by posting ads online with Firetruck memorabilia enthusiast websites. In order to install the truck in the Fireman's Fund Office facility in Alpharetta, Georgia, the truck had to be disassembled by a local Model-T club member and reassembled inside the building. PHOTO CREDIT: Fireman's Fund Insurance Company

1919 Ford Model TT/1914 Howe

The Mowrystown Volunteer Fire Department purchased a new Howe horse-drawn motorized pumper in 1914 which was their only piece of fire apparatus until 1923 when the Department purchased a used 1919 Ford Model TT commercial truck chassis. The forward portion of the 1914 Howe was cut-off with the motorized pump mounted on the Ford Chassis. This unusual combination of a 2-engined pumper was used, coupled with a 2 wheel hose wagon pulled behind as the only pumper until 1947 when the Ford/Howe was replaced. PHOTO CREDIT: Mark Radkte

HOWE FIRE APPARATUS COMPANY

MODERN FIRE FIGHTING EQUIPMENT

Anderson, Indiana, U. S. A.

Howe Fire Apparatus Company letterhead with Ford Model T

PHOTO CREDIT: Steve Hagy Collection

Mark Radtke restored this beautiful 1919 Ford Howe Pumper. The Howe Pump was a horse drawn pumper and the Mowrystown, OH Fire Department placed the pump on a 1919 Model T Ford chassis. PHOTO CREDIT: Mark Radtke.

Ford Model T Childs Pumper Matt Lee Collection

Ford Model T Obenchain Boyer Chassis with pump Matt Lee Collection

Ca 1925 Ford Model T Obenchain Boyer Chemical. Matt Lee Collection

Ford Model T Childs Chemical Car Matt Lee Collection

Ford Model T Foamite Childs Tanker Matt Lee Collection

Ford Model T Prospect Double Tank Chemical

Yorkville's circa 1916 Chemical on car chassis. PHOTO CREDIT: Tod Wirth Collection

1919 Ford Model T Prospect Double Tank Chemical

The Brookville Fire Department, Brookville, Ohio, still owns this 1919 Ford Prospect Model C Chemical & Hose engine. PHOTO CREDIT: Steve Hagy Collection

1919 Ford Model T Prospect Pumper

This unit served Morley, Michigan and later Genera, Michigan. Now privately owned, whether the hose reel originally came with the rig is not known. PHOTO CREDIT: Steve Hagy Collection

1919 Ford Model T Howe Pumper

Recruiting Photograph? Or kids hitching a ride on Rehrersburg, Pennsylvania fire department's Howe pumper? PHOTO CREDIT: Steve Hagy Collection

Circa 1919 Ford Model TT Prospect

This triple combination served Tawas City, Michigan and was photographed in front of the Marion, Ohio fire department headquarters. PHOTO CREDIT: Tod Wirth Collection

1919 Ford Model T American LaFrance Hose and Chemical Combination

The City of Wytheville, Virginia purchased this 1919 Ford Model T American LaFrance Hose and Chemical Combination in late 1919 or early 1920. The Model T remained in active service until 1928 when their 1928 Seagrave pumper was delivered and was loaned to the town maintenance department to be used out on the streets.

During the month of March 1957, Frederic H. Morehead of Wytheville initiated a historical research on the Model T with letters to the American LaFrance Corporation in Elmira, New York and the Ford Motor Company in Dearborn, Michigan, seeking information about the City's 1919 fire engine.

As Ford Motor Company did not produce a fire truck for sale as a commercial item, it was the custom that Ford Motor Company supplied the chassis only to companies such as American La France.

It was quite clear American LaFrance did not have the "full particulars on this piece of equipment" and stated in their March 20, 1957 letter "we find that it is an old Model-T-32 Village Queen shipped from Elmira in 1919."

The Ford Motor Company confirmed the truck had been built in 1919 and that the motor block had been cast on October 2, 1919. The complete responses from American LaFrance and Ford Motor Company follow on the next two pages. PHOTO CREDIT: City of "Wytheville, VA

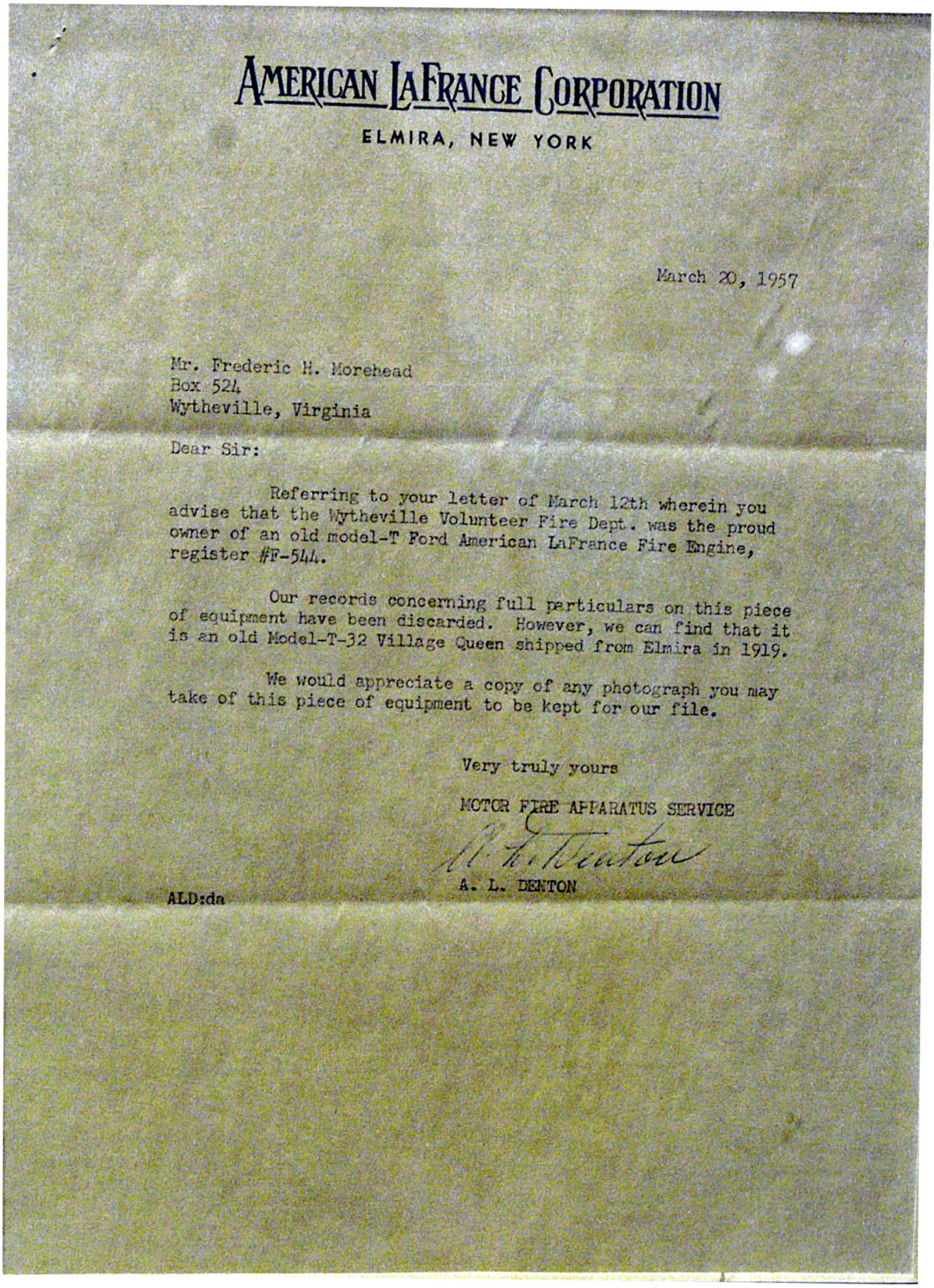

AMERICAN LAFRANCE CORPORATION

ELMIRA, NEW YORK

March 20, 1957

Mr. Frederic H. Morehead
Box 524
Wytheville, Virginia

Dear Sir:

Referring to your letter of March 12th wherein you advise that the Wytheville Volunteer Fire Dept. was the proud owner of an old model-T Ford American LaFrance Fire Engine, register #F-544.

Our records concerning full particulars on this piece of equipment have been discarded. However, we can find that it is an old Model-T-32 Village Queen shipped from Elmira in 1919.

We would appreciate a copy of any photograph you may take of this piece of equipment to be kept for our file.

Very truly yours

MOTOR FIRE APPARATUS SERVICE

A. L. DENTON

ALD:da

American LaFrance letter

American LaFrance response of March 20, 1957 from A. L. Deeton, Motor Fire Apparatus Service. PHOTO CREDIT: City of Wytheville,

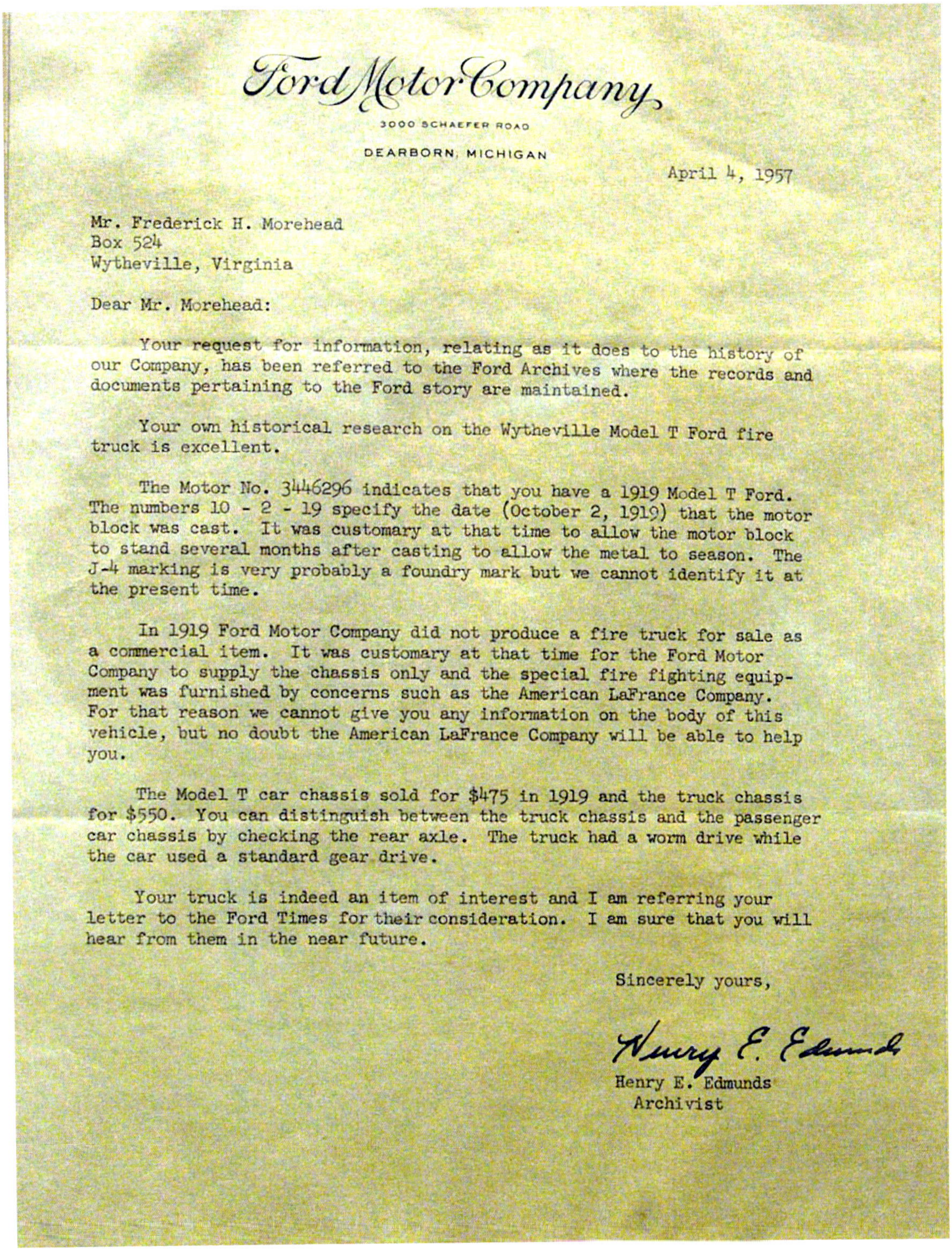

Ford Motor Company

3000 SCHAEFER ROAD

DEARBORN, MICHIGAN

April 4, 1957

Mr. Frederick H. Morehead
Box 524
Wytheville, Virginia

Dear Mr. Morehead:

Your request for information, relating as it does to the history of our Company, has been referred to the Ford Archives where the records and documents pertaining to the Ford story are maintained.

Your own historical research on the Wytheville Model T Ford fire truck is excellent.

The Motor No. 3446296 indicates that you have a 1919 Model T Ford. The numbers 10 - 2 - 19 specify the date (October 2, 1919) that the motor block was cast. It was customary at that time to allow the motor block to stand several months after casting to allow the metal to season. The J-4 marking is very probably a foundry mark but we cannot identify it at the present time.

In 1919 Ford Motor Company did not produce a fire truck for sale as a commercial item. It was customary at that time for the Ford Motor Company to supply the chassis only and the special fire fighting equipment was furnished by concerns such as the American LaFrance Company. For that reason we cannot give you any information on the body of this vehicle, but no doubt the American LaFrance Company will be able to help you.

The Model T car chassis sold for $475 in 1919 and the truck chassis for $550. You can distinguish between the truck chassis and the passenger car chassis by checking the rear axle. The truck had a worm drive while the car used a standard gear drive.

Your truck is indeed an item of interest and I am referring your letter to the Ford Times for their consideration. I am sure that you will hear from them in the near future.

Sincerely yours,

Henry E. Edmunds

Henry E. Edmunds
Archivist

Ford Motor Company letter

Ford Motor Company archivist Henry E. Edmunds' response of April 4, 1957. PHOTO CREDIT: City of Wytheville,

1919 Ford Model T American LaFrance Hose and Chemical Combination

Restoration of Wytheville's Model T began in the fall of 1995 and completed in the spring of 1997.

1919 Waterous Ford Model T Pumper, Frankenmuth, MI

PHOTO CREDIT: Walt McCall

SPECIFICATIONS—Continued

EXTINGUISHER—One AJAX "approved" hand fire extinguisher, in polished copper.

EXTINGUISHER HOLDER—One brass, mounted on running-board.

LADDER—One 16 ft. extension ladder complete with suitable holders.

TAIL LIGHT—One oil.

HEAD LIGHTS—Two electric.

SIDE LIGHTS—Two oil.

WRENCHES—All necessary wrenches.

PIPING—Brass.

TOOL BOX—One mounted on running-board.

PRESSURE GAUGES—Two.

JACK—One.

TIRE PUMP—One.

TOOLS—Complete set motor and chassis.

TIRE REPAIR KIT—Complete.

PAINTING—Entire car Red, Black and Gold striped, with all brass parts highly polished.

AJAX Combination Chemical and Hose Car.

FOR use in Towns which have water-works, a Combination Chemical Engine and Hose Car is frequently preferred to the Double Tank Chemical Car described on the previous page. The Combination Car, described below, carries only one chemical tank of 35 gallons, but it has in addition a body providing capacity for 500 feet of ordinary water fire hose. With the exception of the fact that it has only one tank, instead of two tanks, it is just as efficient a chemical engine as the car described on the previous page. An ideal Combination Chemical and Hose Car for Towns having the advantage of water-works.

SPECIFICATIONS

CHASSIS—Ford Motor Company's standard chassis, 4 cylinder, 3¾ in. x 4 in., 20 H. P.

HOSE BODY—Has a capacity for 500 feet of double jacket hose, or any similar hose. This water hose is not included in our equipment, as most Towns already have their own water hose. The chemical hose in the hose basket is furnished by us, however.

CHEMICAL TANK — One, 35 gallons capacity, made of heavy steel, thoroughly protected. Capable of standing more than 10 times the average working pressure.

CHEMICAL HOSE—150 feet ¾ in. four-ply rubber chemical hose with heavy brass couplings attached.

BASKET—One wire hose basket.

NOZZLE—One brass shut-off nozzle.

ACID RECEPTACLES—Two.

ACID RECEPTACLE HOLDER—One, mounted on running-board.

CROW-BAR—One, fire-department style, with proper holder.

AXE—One, fire-department style, with proper holder.

LANTERNS—Two, fire-department style, with proper holders.

SODA CANISTER—One, mounted on running-board

EXTINGUISHER—One AJAX "approved" hand fire extinguisher.

EXTINGUISHER HOLDER—One, mounted on running-board.

LADDER—One, 16 ft. extension ladder.

TAIL LIGHT—One oil.

HEAD LIGHTS—Two electric.

SIDE LIGHTS—Two oil.

WRENCHES—All necessary wrenches.

PIPING—Brass.

TOOL BOX—One, mounted on running-board.

PRESSURE GAUGE—One.

JACK—One.

TIRE PUMP—One.

TOOLS—Complete set motor and chassis.

TIRE REPAIR KIT—Complete.

PAINTING—Entire car Red, Black and Gold striped, with all brass parts highly polished.

Pages 1 and 3 of an Ajax Sales Flyer circa 1920's. Steve Hagy Collection

1920 Ford Model T Prospect Double Tank Chemical

Chip Lytle's 1920 Ford Prospect, lettered "D.F.D." has two 30 gallon Deluge chemical tanks. PHOTO CREDIT: Tod Wirth Collection

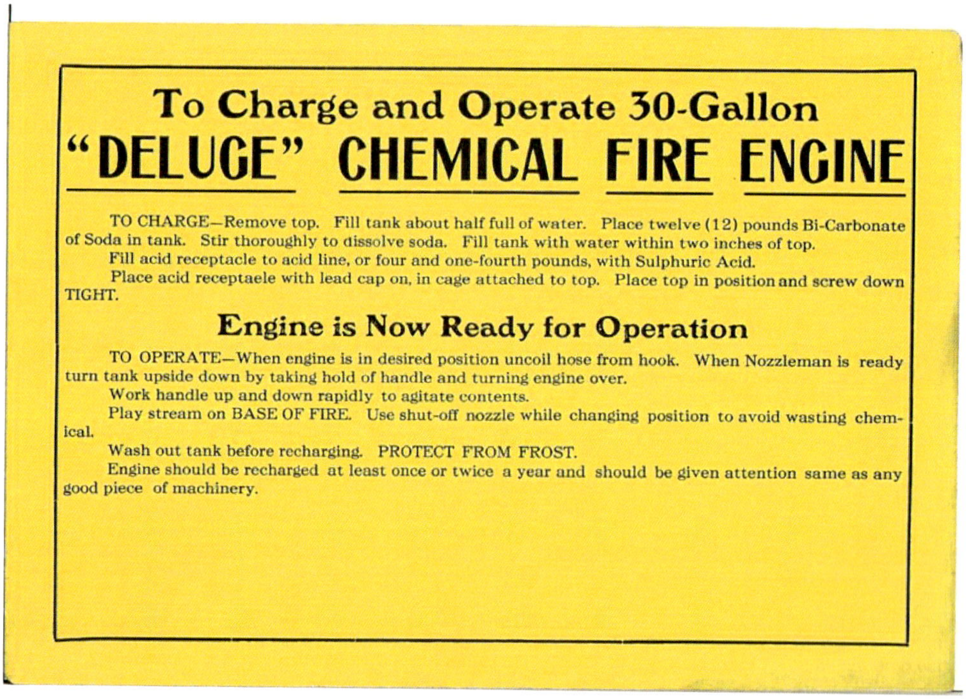

To Charge and Operate 30-Gallon

"DELUGE" CHEMICAL FIRE ENGINE

TO CHARGE—Remove top. Fill tank about half full of water. Place twelve (12) pounds Bi-Carbonate of Soda in tank. Stir thoroughly to dissolve soda. Fill tank with water within two inches of top.

Fill acid receptacle to acid line, or four and one-fourth pounds, with Sulphuric Acid.

Place acid receptaele with lead cap on, in cage attached to top. Place top in position and screw down TIGHT.

Engine is Now Ready for Operation

TO OPERATE—When engine is in desired position uncoil hose from hook. When Nozzleman is ready turn tank upside down by taking hold of handle and turning engine over.

Work handle up and down rapidly to agitate contents.

Play stream on BASE OF FIRE. Use shut-off nozzle while changing position to avoid wasting chemical.

Wash out tank before recharging. PROTECT FROM FROST.

Engine should be recharged at least once or twice a year and should be given attention same as any good piece of machinery.

1920 Ford Model T Prospect Deluge Chemical Fire Engine Instructions

Detailed instructions to charge and operate the "Deluge" Chemical Fire Engine. PHOTO CREDIT: Tod Wirth Collection

Eaton (CO) Fire Department's 1920 Ford Model T Chemical Engine

The Eaton fire department was established January 1, 1914. The fire department archivist believes this is Eaton's 1920 Ford Model T Chemical engine. This photo was reproduced from old negatives from bout 1920 recently discovered n their old station. PHOTO CREDIT: Eaton Fire Department

19?? Ford Model T American LaFrance Double Tank Chemical

This unit served Amelia Ohio fire department. PHOTO CREDIT: Steve Hagy Collection

LUVERNE FORD TRIPLE COMBINATION

The logical Fire Truck for small towns and for auxiliary equipment
or suburban service in large towns

Right Side View

A Properly Balanced Combination
of the
FORD TRUCK CHASSIS
and the
LUVERNE
Fire Apparatus Equipment

A Complete
PUMPER, CHEMICAL HOSE
and LADDER TRUCK
of Light Weight, Low Cost and
Great Efficiency

Left Side View

DETAILED DESCRIPTION OF THE LUVERNE FORD FIRE TRUCK EQUIPMENT

HOSE BODY—Made of 14 gauge steel sheets, attached to a wood frame base. Upper edges of body are re-inforced with angle bars. Ladder carriers are attached to body. Hangers for gasoline tank at front end.
Capacity of body, 800 ft. of hose.

DRIVER'S SEAT—Made of wood, and attached to front end of body. Upholstered with heavy artificial leather.

REAR FENDERS—Heavy steel, attached to body and connected to side and rear step.

RUNNING BOARDS—Full length side running boards and a substantial rear step, all carried by hangers from body sills.

PUMP—Rotary type 200 gal. capacity. To mount under the driver's seat and connect direct to the Ford transmission. Has pressure gauges, relief valve and cooling line for connecting to the Ford radiator.

CHEMICAL TANK—One Champion Steel Tank of 35 gallons capacity, mounted on top of hose body.
Heavily tinned inside and outside.

BY-PASS PIPING—Best grade brass piping, with pressure gauge and booster line to pump.

CHEMICAL HOSE—150 feet of best grade ¾-inch 4-ply hose, with one shut off nozzle, carried in a woven wire hose basket mounted over the hose body.

LADDERS—One 12-foot Roof Ladder, with folding hooks, one 20-foot rope hoist Extension ladder. Both ladders are carried on right side of truck.

MISCELLANEOUS—Two 2½-gallon Phomene Extinguishers with charges, Two Oil Lanterns, 1 Bell, 1 Axe, 1 Crowbar, 1 Pikepole, One set By pass Piping for connecting to Chemical Tank and to pump, 2 Acid Receptacles, 1 Soda Canister, 1 Acid Canister, Two 10-foot lengths of 3-inch Stiff Suction Hose, with Strainer and Adapter, all necessary spanners and wrenches, Two Nozzle Standards.

GUARANTEE—All parts of the Equipment are guaranteed for a period of one year.

FINISH—Painted in a high grade manner with Fire Department red, and gold leaf striping. All fittings nickel plated.

CRATING—All parts are well crated for shipment anywhere in the U. S.
An extra charge is made for export crating.

PRICE FOR COMPLETE SET OF EQUIPMENT AS DESCRIBED $2200.00

The Equipment is fully painted and finished, ready to mount onto a Ford Chassis.
We furnish full instructions, so that anyone can mount the equipment, or if the chassis is sent to our factory we will do the mounting for $40.00.

LUVERNE FIRE APPARATUS CO., Luverne, Minn.

Manufacturers.

NORTHWESTERN FIRE SUPPLY CO., Fargo, No. Dak., Distributors

Luverne advertisement from the Steve Hagy Collection

1920 Luverne Ford Model T Pumper

Ex-Ellsworth, Minnesota Luverne Ford Model T pumper owned by Neil and Mary Ann O'Brien, Albuquerque, NM. PHOTO CREDIT: Walt McCall Collection

1920 Ford Model T Luverne Pumper

Luverne advertising demonstration of Ellsworth, Minnesota's1920 Ford Model T Luverne pumper. PHOTO CREDIT: Neil O'Brien

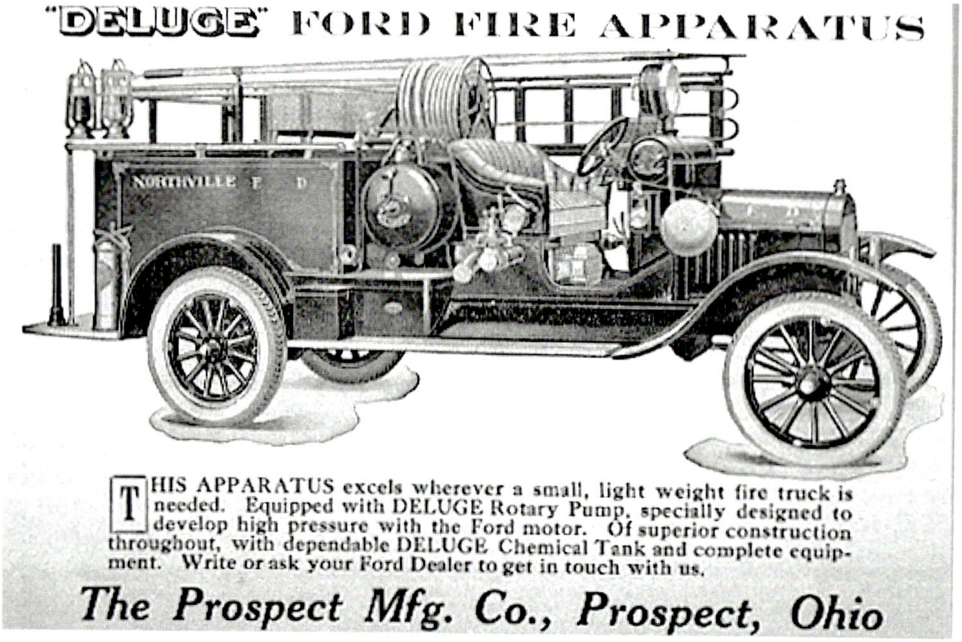

Prospect Mfg. Company advertisement from the Steve Hagy Collection

1920 Prospect Deluge Chemical

Arlington, Ohio Volunteer Fire Department. PHOTO CREDIT: Bill Killen Collection

Howe-Ford D.T. Double Combination Chemical and Hose on Ford One Ton Chassis

Howe Manufacturing Company Advertisement

This Howe advertisement featured the Dallas City, Illinois fire department's combination chemical and hose on a Ford One Ton Chassis. PHOTO CREDIT: Steve Hagy Collection

Ford Model TT Howe Double Tank Chemical and Hose

Dallas City, Illinois fire department operated this combination double tank chemical and hose on a Ford One Ton Chassis. PHOTO CREDIT: Steve Hagy Collection

New Washington, Ohio

In the summer of 1920 Fire Chief Pessefall lobbied for a siren for the town and a new piece of fire apparatus that would be capable and faster at responding to fires than their current hand and horse drawn equipment. This led to the village and the township jointly purchasing a siren and a 1920 Ford Model T Prospect Model C Chemical engine.

The Village of New Washington and Cranberry Township executed a purchase order to the Prospect Manufacturing Company for a Prospect Model C fire engine on a Ford Model T chassis for $1,622.49. Records indicate that when time came to pay for the engine the price was rounded down to $1,600.

1920 Ford Model T Prospect Model C Chemical engine

The Mayor of New Washington, Ohio and the Fire Chief of the Cranberry Township fire department take the new fire engine for a drive. PHOTO CREDIT: Tod Wirth Collection

"Small Blaze

The fire department was called out Monday (December 13) evening to the Moorhead flat above the Pessefall hardware to extinguish a blaze caused by a coal oil stove. The fire was soon extinguished with the new chemical engine. The loss was light.

The chemical system proved to be a great factor in fires of the above kind and means much for fire protection in this community." - New Washington Herald, Friday December 17, 1920

Interesting Note

Pessefall Hardware was owned and operated by the Fire Chief

1920 Ford Model T Prospect Model C Chemical engine

Fire Chief Pessefall with the Cranberry Fire Department's Prospect Model C chemical engine which appears to have been in service for some time. PHOTO CREDIT: Tod Wirth Collection

1920 Ford Model T Prospect Model C Chemical engine

The Cranberry Fire Department's Prospect Model C chemical engine as restored by the Tubbs Insurance Company, now owned by Tod Wirth. PHOTO CREDIT: Tod Wirth Collection

1920 Ford Model T Emergency Tender

This 1920 Ford Model T Emergency Tender served in Gentofte, Denmark and was featured in Enjine! Enjine!. PHOTO CREDIT: SPAAMFAA ENJINE! ENJINE!

.

Low in Cost

High in Value

DELUGE FORD CHEMICAL CAR

Advertisement Prospect Manufacturing Company, Prospect, Ohio

City of East Detroit Fire Department's 1921 Ford Model T Engine

This Model T Ford Fire Engine, serial No. 6413779, originally purchased by the City of East Detroit at a cost $1,802.96 including equipment. This Ford Chassis is one of 5255 units built August 18, 1922. PHOTO CREDIT: Detroit Historical Museum

1921 Waterous Ford Model T Pumper

Waterous built this rear mounted 300 GPM Rotary Gear pumper for Glenwood City, Wisconsin. PHOTO CREDIT: Walt Schryver ca 1975

1921 Ford Model T Double Tank Chemical

Lettered "Half Way F.D.", this is East Detroit's recently restored rig maintained in the Detroit Historical Museum. PHOTO CREDIT: Nick Sage.

1921 Ford Model T Chemical

Indian River fire department's Ford chemical is unique in that the rear wheels are solid rubber and the head lights are circa 1914, according to my read of "Bruce McCalley's book, "Model T Ford". PHOTO CREDIT: Tom Herman Collection

The following pictures and commentary were provided by owner Dave Mohr, West Chester, Ohio

1921 Ford Model T American Front Mount Pump

"I am clueless as to its origin, manufacturer or service history. I do know that it was originally a chemical truck converted to a water pumper, possibly in the forties. It has a water tank where the chemical tanks would have been, and an American front mount pump.

The truck was purchased some time in the 1950's as a toy and never saw service in Silverton. The old guys that knew more of the history of the truck are no longer among the living. The truck got stashed in a barn and forgotten about. No one had interest in it and when the barn and property were auctioned off I inherited it. I had been aware of the truck since I was a wee lad, but thought it was long gone. One story has it being purchased from a junkyard in Tennessee; another has it coming from somewhere in Georgia".

-- Dave Mohr, West Chester, Ohio

"I just found some info on the 1927 Model T. The unit was bought from a junk yard in London, Ohio in 1953 for $100.00. The first owner of the unit was Grove Port, Ohio, and the second owner was Westerville, Ohio. It was a chemical truck and then the pump was added at a later date. The fire pump was an American Marsh Pump made by American Steam Pump out of Chicago, Ill The tank was off an old Ahrens Fox that was found in a junk yard in Sharonville, Ohio. The spot and red light came off the same unit."

Chief Don Newman, Deer Park Silverton Joint Fire Department.

As this book went to press research was still on-going to finalize the actual date the truck was manufactured.

1921 Ford Model T American Front Mount Pump and Silverton Fire Department's 1991 Seagrave Rescue Pumper

Dave Mohr's co-workers arrived in Silverton's Engine 89 to assist in off-loading Dave's Model T fire truck May 1, 2007. PHOTO CREDIT: Dave Mohr

1921 Ford Model T Double Tank Chemical

Commack Hook and Ladder Company No. 1, Commack New York operated this double tank chemical. PHOTO CREDIT: Tom Herman Collection

Tally Ho Engine Company No.3

1921 Ford Model T Chemical Engine

PHOTO CREDIT: Mike Chiaramonte

"Tally Ho Engine Company No.3 was organized in 1912 and merged with the Lynbrook, NY fire department in 1922. The Tally Ho Fire Company Model T Ford firetruck was made in 1921 as a one ton truck chassis. A carriage builder in Troy New York created its body from the dashboard to the rear step. It is a chemical truck carrying two 40 gallon tanks. Soda and acid were mixed with the water in the tanks to create gas that expelled the water. In it's operating days it carried an additional 160 gallons of water in 16 ten gallon milk cans in the hose bed.

The truck served the Upper Jay N.Y. Fire Dept. from 1921 to 1965. In its last decade of service it was used as a utility truck (carrying beer kegs, tables and chairs to picnics). During the fifteen years between '65 and '80 it sat outdoors at the North Pole Village, a children's amusement park in North Pole ,N.Y. In 1980 while at a firemen's convention in Lake Placid, NY the men of Tally Ho Eng. Co. 3 of the Lynbrook NY fire dept. discovered it moldering away behind a gas station.

The truck was a duplicate of one that served Tally Ho in the 20's. All the wood, leather and rubber were rotted away. The men bought it for $1,800, raised an additional $1,800 and began restoration. We incorporated an Antique Truck Company and Mike Chiaramonte served at its first Chairman. All the wooden parts were faithfully recreated using the exact kind of wood as the original. Ford parts were scavenged from antique auto parts dealers, swap meets, auto flea markets and musters. Some parts were made to match the originals. The sale of wooden nickels and a $5,000 Federal Grant helped to complete the task of bringing it back to life.

In 1983 it was painted red, buffed bright and shiny and purring like a kitten. The past 15 years it has been our pride and joy and the most photographed entrant in every parade we attend. We're justifiably proud of "Old Jubilee"

- Jim Grismer.

Big Four Fire Company

On October 4, 1920 a group of citizens from the Pennsylvania communities of Lampeter, New Danville, Refton and West Willow met to discuss, after considerable agitation, the subject of fire fighting apparatus for the communities.

A meeting, held at the Lampeter High School, was called to order by the temporary chairman, Benjamin L. Herr. After lengthy discussion a Board of Directors was elected, and the first meeting adjourned after adopting "The Big Four Fire Company" as the official name.

Over the next several months funds were raised through donations and pledges to order four fire engines. On January 17, 1921 the Big Four Fire Company awarded a contract to the Prospect Manufacturing Company of Prospect, Ohio in the amount of $7,339.00 for four complete units to be delivered to Lancaster.

Big Four Fire Company's 1921 Ford Model T Prospect Chemical Engines

This fleet of Ford Model T Prospect fire engines were photographed in front of the North and South buildings of the Prospect Manufacturing Company, Prospect, Ohio March, 1921 prior to delivery in Pennsylvania. PHOTO CREDIT: Tod Wirth Collection

The new chemical fire engines were built on the Model TT truck chassis and carried two 60 gallon chemical tanks and a single hose reel. The engines arrived in Lancaster on April 1, 1921 and were turned over to the Big Four Fire Company on April 7 after a final demonstration before more than 200 citizens at the Willow Street Vocational School.

Mr. Paul Graybill was the Fire Chief of Refton and his salary was set at $5.00 per month.

On May 18, 1922 it was approved to place and maintain a fire machine East end of Strasburg Borough; and on December 4, 1922 the Township Supervisors reported to the Big Four Fire Company that they had levied a 1/2 mill tax for maintenance of the Big Four Fire Company's fire apparatus.

Records of fires where Ford Model T fire apparatus responded and operated are rather sparse and photographs of Ford Model T fire engines in operation on the fire ground are rarer still. Some of the more interesting fires responded to by the Big Four Fire Company reveal the success and innovation of firefighters of the "roaring twenties".

Big Four Fire Calls

One of the first responses after the Prospect Chemical engines were delivered occurred on April 11, 1921 was to the Toll House at New Danville Pike. The fire was out when the engine arrived within 10 minutes from fire house.

Three engines from Lampeter, Refton and West Willow saved Mr. Cowden's summer house near Strasburg on October 15, 1921 and on October 30, 1921 the Refton and West Willow engine companies saved Mr. Smith's house at the Union valued at $1,500.

The Lampeter School was destroyed on March 22, 1922, with all Big Four Fire Company engines on the scene in good time. Adjoining property was protected.

All four engines responded to the J. H. Hildebrand butcher shop at Willow Street on March 1, 1923. The adjacent property saved.

March 30, 1923. The Refton engine was called to put out a fire on the roof of George Miller's barn. The fire was caused from sparks from low grade engine. The building was saved, with the roof partially destroyed. ***An interesting aspect in connection with this call worthy of mention: A spark plug fouled on the way. Mr. Ed Evans, the mechanic, put a new plug in while on the way without stopping the fire engine, displaying courage and nerve***.

Imagine changing the plug on this engine while responding to a fire.

As the individual fire companies grew the Big Four Fire Company was disbanded on February 27, 1929. PHOTO CREDIT: Refton Community Fire Company

1922 Ford Model T Bickle Two Tank Chemical Car

Bickle built this two tank chemical car for the Waterville-Cambridge Fire District, Ontario Canada. PHOTO CREDIT: National Archives of Canada/Walt McCall Collection

1922 Ford Model T Boyer Triple Tank Chemical Car

Boyer built this triple tank chemical car for the Bronson, Michigan fire department. Each chemical tank held 35 gallons of agent. PHOTO CREDIT: Rodger Birchfield Collection

1920's Hale Ford Model T Single Tank Chemical Car

Hale built this single tank chemical car for the Croydon, Pennsylvania Fire Company No. 1 in the 1920's. PHOTO CREDIT: Walt McCall Collection

1920's Hale Ford Model T Hose and Chemical Combination

This Hale hose and chemical combination was delivered to the Repaupo, New Jersey fire department in the 1920's. PHOTO CREDIT: Walt McCall Collection

1922 Ford Model T American LaFrance Double Tank Chemical Car

This American LaFrance double tank chemical car served the Squirrel Hill, Pennsylvania fire department. PHOTO CREDIT: Walt McCall

1922 Ford Model T American LaFrance Fire Chief's Runabout

This Fire Chief's Runabout, originally delivered as an American LaFrance Chemical car Registration Number F 820, served fire departments in Michigan and Ohio. The chemical tanks were removed in the 1930's and replaced with an equipment box. F. Parker Browne purchased the car in the mid 1980's and restored it to its present condition and sold it Phil & Dave McLean, Richfield, Ohio. PHOTO CREDIT: Phil McLean.

1922 Ford Model T Hose Wagon

This 1922 Ford Model T Equipment Carrier was locally built and photographed at the 2006 SPAAMFAA National Muster in Columbus, Ohio. PHOTO CREDIT: Steve Hagy Collection

Ford Model T Double Tank Chemical

Ca 1923 Ford Model T double tank chemical operated by the Chicago Fire Department, manufacturer unknown. PHOTO CREDIT: Tom Herman Collection

1922 Ford Model T Double Tank Chemical

This Ford Model T double tank chemical was used in New Portland, Maine and is displayed courtesy of Andrew Wilson Edmonds. PHOTO CREDIT: Tom Herman Collection

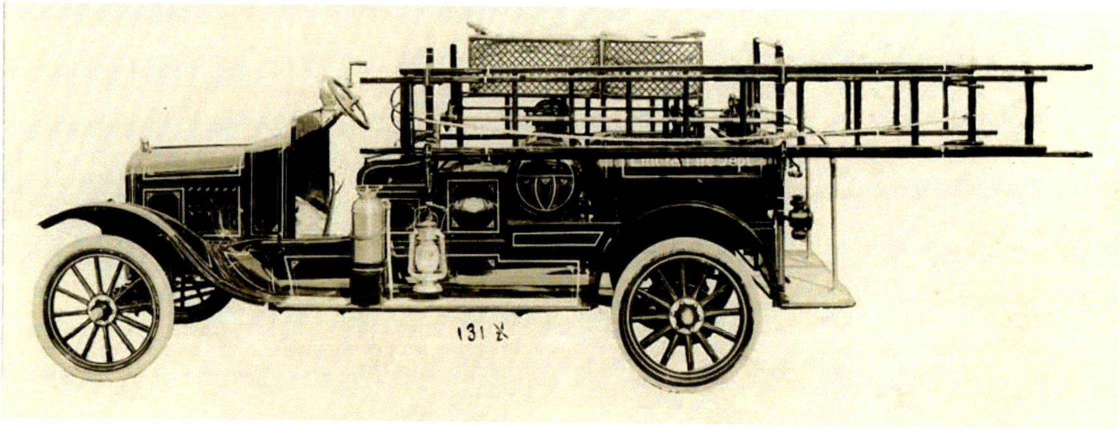

1922 Ford Model T Boyer Triple Chemical

Factory photograph of Elnora, Indiana Fire Department's 1922 Ford Model T Boyer three tank chemical. PHOTO CREDIT: Rodger Birchfield Collection

1922 American Fire Apparatus Ford Model T Front Mount

The first Barton pump was mounted on this 1922 Ford Model T by American Fire Apparatus, nicknamed "Thunderbolt" . PHOTO CREDIT: Tom Herman Collection

1922 Brockway Ford Model T

The Rockville (Maryland) Volunteer Fire Department's 1922 Brockway had a rear mount pump, located behind the right rear wheel. Note the chains on the rear tires. PHOTO CREDIT: Russell Dawson

1922 Ford Model T American LaFrance Chemical & Hose

Farmland (Indiana) Volunteer Fire Department's 1922 Ford Model T American LaFrance Double Tank Chemical and hose was built on the one ton Model TT chassis. PHOTO CREDIT: Steve Hagy Collection

1922 Ford Model TT Waterous #944 Pumper

St Clair Fire Protection District purchased this rig from Benld, Illinois fire department in 1931. Saw service with the Harrisonville-Legionnaire fire department where the pump was removed and secured by St Clair and underwent restoration in 1991. PHOTO CREDIT: Dennis Maag.

1922 Ford Model T Waterous Matt Lee Collection

1922 Ford Model T Hose Car Bill Killen Collection

1922 Ford Model TT Double Tank Chemical

Hot Lake Sanitarium operated this double tank chemical in Eastern Oregon. Owned by the LaGrande fire department, it is exhibited in the Eastern Oregon Fire Museum. PHOTO CREDIT: Leon Brewer.

"This 1922 Ford, Model "TT" Fire Truck was purchased by Don Keeling Asst. Chief of La Grande Volunteer Fire Department at a public auction on October 13, 1999. The truck was originally purchased for the fire protection of Hot Lake Sanatorium (Hot Lake is an historic sanatorium about 9 miles southeast of La Grande).

"The two tanks on the back were filled with a mixture of bicarbonate soda and water. Upon arriving at a fire a lever on the tank would be rotated pouring a vial of sulfuric acid within the tank into the mixture causing a violent chemical reaction capable of producing up to 400 lbs. pressure within the vessel.

The mixture under pressure would be directed through the chemical hose above the tank and applied to the fire to put it out. Items missing from the truck include two lanterns that were mounted on the upper rear of the truck (brackets are still there), spotlight and bell originally mounted on the cowl, covers for the soda acid tanks and the levers to operate the acid mixture, the seat and other minor items. During restoration, a new seat was installed."--

Kathleen Almquist
Eastern Oregon Fire Museum & Learning Center

Clayton, NY's 1918 Ford Model T Fire Chief's Car

Former Clayton Fire Chief Art Benett enjoys a 4th of July drive in Clayton Fire Department's 1918 combination Chief's buggy and hose wagon. PHOTO CREDIT: William Bender Collection, Thousand Island Museum, Clayton, NY

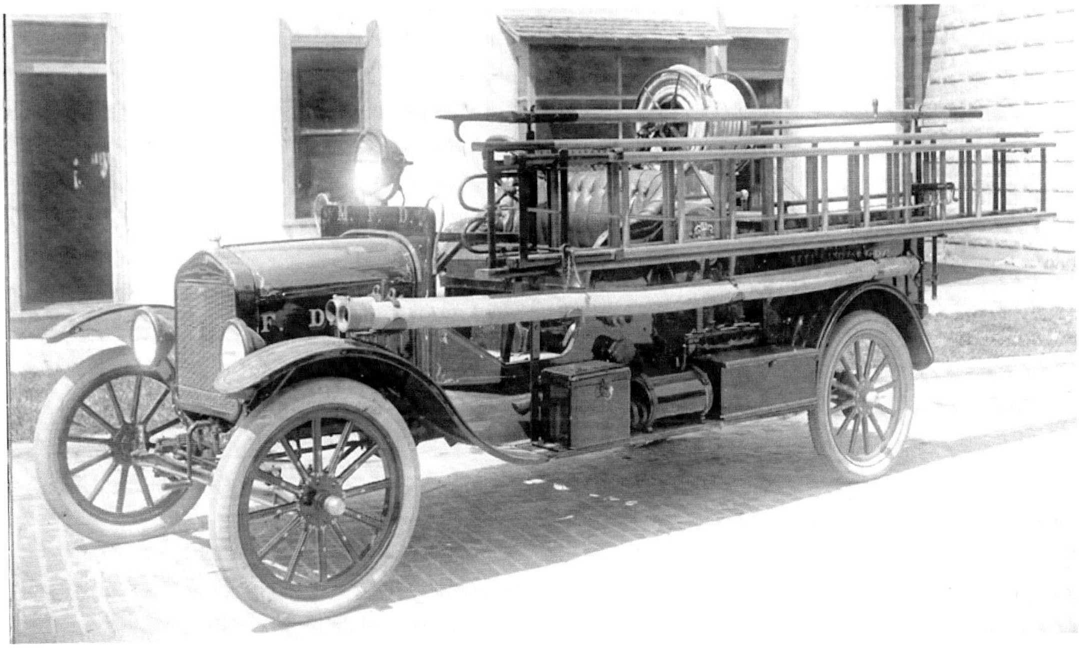

1922 Ford Model T Prospect Triple Combination

This Prospect Triple Combination was built on an extended passenger car chassis for a Michigan fire department. PHOTO CREDIT: Tod Wirth Collection

1922 Ford Model T Howe Pumper

The Takoma Park, (Maryland) Volunteer Fire Department, a suburb of Washington, DC operated this 1922 Ford Model T Howe pumper. PHOTO CREDIT: Matt Lee Collection

Los Angeles County, California's 1923 Ford Model TT Chemical Hose Car

This chemical/hose car has been part of the Los Angeles County Fire Department since the early 1960s, when it was designated as Engine 1/2. Powered by the Ford four-cylinder Model T engine, this vehicle is larger than its 1915 predecessor. Engine 1/2 is often used in public fire safety education campaigns. PHOTO CREDIT: County of Los Angeles Fire Museum Collection

Toronto Fire Department, Ontario, Canada's 1923 Ford Model T Chief's Runabout

The Toronto Fire Chief's 1923 Ford Model T Runabout, at Fire Station 19 on Perth Avenue, Toronto with a 1919 American LaFrance pumper. PHOTO CREDIT: Walt McCall Collection

1923 Ford Model T Howe Double Tank Chemical

Vermontville, Michigan fire department operated this Model T Howe double tank chemical engine. PHOTO CREDIT: Steve Hagy Collection

Middleton Fire Department's 1923 Ford Model T Obenchain-Boyer

Here is a photo of the 1923 Model TT fire truck restored for the Middleton, WI Fire Department. It took about a year to restore and was finished in early 2007. The truck manufacturer is Obenchain-Boyer. Middleton Fire Department was the original purchaser.

1923 Ford Model T Howe Pumper

Ashland, Illinois fire department's 1923 Ford Model T Howe 250 GPM pumper with pre-connected squirrel tail hard sleeve. PHOTO CREDIT: Ron Manwaring

1923 Ford Model T Prospect

1923 Ford Model T Prospect

How long the Dry Ridge Kentucky fire department owned and operated this Ford Model T Prospect is unknown. The current owner, Richard Willoby of Williamstown, Kentucky, is in the process of restoring the rig. PHOTO CREDIT: Jerry Sudduth Collection

1923 Ford Model TT Howe Triple Combination

"The 1923 Ford Howe was in the Deer Lodge, Montana Towe Museum for a long time, but I am not sure how long. I think it was around 30 years, and then the Towe found itself in debt because of IRS actions so they auctioned off a number of vehicles including the 1923 Ford Model T Howe. I am not sure who purchased it, but I understand he was a multi millionaire who lived in the Palm Springs CA area. He took the 23 Howe to his home and parked it in front as a "lawn ornament." After he got tired of it he sold it to a fellow who lived in San Clemente and was a Battalion Chief at the nuclear power plant fire department south of San Clemente. He tried unsuccessfully to restore the vehicle, and finally got frustrated and sold it to me. When the new owner purchased it from the Towe Museum, it was lettered "Bowen". I have not been able to find additional information on this truck and the only "Bowen" I found is in Illinois." – Wes Melo
PHOTO CREDIT: Wes Melo

1923 Ford Model T Chemical and Hose

The manufacturer of this combination chemical and hose wagon owned by the Swanton Fire Department, Swanton, Ohio is unknown. PHOTO CREDIT: Bill Friedrich Collection

1923 Ford Model T

Novato, California's first motorized fire engine was this Model T passenger car converted to civic use by adding a fire pump, hose and chemical tank. Fireman's Fund Insurance Company acquired this unit in 1984 and restored it. It is currently on display at the Home Office. PHOTO CREDIT: Fireman's Fund Insurance Company

1923 Ford Model TT Prospect Model E B Pumper

Monroe, Ohio's first motorized fire apparatus was this 250 GPM Prospect pumper with a 60 gallon chemical tank. PHOTO CREDIT: Steve Hagy Collection.

1923 Prospect Ford Model T Chemical Car

The Cedar Point Amusement Park, Sandusky, Ohio, used this second generation Model C Chemical, delivery photo above, and restoration photo below. PHOTO CREDIT: Stewart Warner

County of Los Angeles Fire Museum Association's Model T's

U. S. Naval Academy's 1924 Model T American LaFrance Chemical Car

This Model T American LaFrance Chemical car was delivered to the U.S. Naval Academy, Annapolis, Maryland in 1924 by the Colonial Motor Company of Annapolis, Maryland. The American LaFrance registry number is F-873. CREDIT: Doug Thomas Retired Fire Marshal Program Administrator.

1923 Ford Model T Chemical and Hose

The manufacturer of this combination chemical and hose wagon owned by the Frankfort Fire Protection District, Frankfort, Illinois is unknown. PHOTO CREDIT: Bill Friedrich Collection

1924 Ford Model T American LaFrance Double Tank Chemical and Hose

Trevarno, California's Double Tank Chemical was part of the Figgie collection when this photo was taken and is currently a part of the North Charleston and American LaFrance museum collection. The Trevarno fire department purchased the chemical tanks and equipment from American LaFrance and installed the equipment on their Model T. PHOTO CREDIT: Dan Martin

1924 Ford Model T Prospect

Rock Creek Fire Department, Rock Creek, Ohio used this 1924 Ford Model T Prospect Combination Hose and Chemical car. PHOTO CREDIT: Walt McCall

1924 Ford Model T Double Tank Chemical

The Village of Rothesay Fire Department was formed in 1924 with all volunteers and their first apparatus was a 1924 Ford Model T chemical unit. The unit pictured was constructed as a bi-centennial project in 1984 with several parts from the original fire apparatus. The Village of Rothesay fire department was incorporated with the Kennebecasis Valley fire department in 1966. PHOTO CREDIT: Bill Killen Collection

1924 Ford Model T Pumper

Greenwood Fire Company's (DE) first motorized fire apparatus placed in service 1924 is equipped with a Hale pump. PHOTO CREDIT: Greenwood Fire Company

1924 Ford Model T Boyer Pumper

This Ford Model T Boyer is the first engine built with a Barton pump. This 150 GPM front mount pumper served the Litchfield, Michigan fire department. PHOTO CREDIT: Robert Rohrback

1924 Ford Model T with American LaFrance equipment

Smyrna, Tennessee fire department's first fire engine cost $990.00 and was equipped with three 50 gallon soda acid tanks. PHOTO CREDIT: City of Smyrna, Tennessee

1924 Bickle Ford Model T Two Tank Chemical Car

This two tank chemical served the town of Madoc Fire Department, Ontario, Canada. PHOTO CREDIT: Walt McCall Collection

1925 Ford Model TT American LaFrance Pumper

This Model TT American LaFrance pumper served Annadale, NJ for several years and was donated to the Annadale Reform School in the 1960's where it served for a number of years. This rig is unusual in that the Ford Planetary transmission was replaced with a Hemco transmission and is part of the Model T Ford Club of America's Museum exhibit. CREDIT: Bill Killen

1925 Boyer Ford Model T Triple Combination Pumper

This 1925 Boyer Model T Triple Combination Pumper was built for the Clarkston, Michigan fire department. PHOTO CREDIT: Walt McCall

Nesco Volunteer Fire Department's 1925 Ford Model T American LaFrance

At 2 AM Sunday morning January 5, 2003 a passerby spotted smoke coming out of the Nesco Volunteer Fire Department building in Mullica Township, New Jersey. By the time firefighters arrived, half of the building was engulfed in flames. The Nesco Volunteer Fire Department dates back to 1925. Their first piece of fire equipment, an American LaFrance Double Tank Chemical, purchased in 1925 for $2250 was lost in the fire. Refurbished in 1952, the double tank chemical had been used in fire prevention activities and won many awards at fire apparatus musters and firefighting shows. In addition to the Model T, a tanker pumper and the fire chief's car with all his turn-out equipment also went up in flames. The last addition to the building was constructed in the 70s. PHOTO CREDIT: Scott Mattson

Nesco VFD's 1925 Ford Model T American LaFrance

1925 American LaFrance Ford Model T Front Mount Pumper

This 1925 American LaFrance Ford Model T front-mount pumper equipped with a Fyr-Foe pump served the Port Hope, Michigan fire department. PHOTO CREDITS: Walt McCall

1925 Ford Model T

This 350 GPM pumper is located in Richmond, Texas. PHOTO CREDIT: Dennis Maag Collection

1925 Ford Model T Double Tank Chemical

This was the first fire truck in the town of Bergholz, New York. PHOTO CREDIT: Phil Gilson

1912 Howe/1925 Ford Model T

The Village of Cecil, Wisconsin's first motorized fire engine was this 1912 Howe pump mounted on a 1925 Ford Model TT chassis. The Village of Cecil was just one of many small communities to transfer fire pumps from horse drawn fire apparatus to the venerable Ford Model T. The Waterous purchase order for the pump is dated June 1912. PHOTO CREDIT: Ronald Czaplicki

1925 Ford Model T Runabout Fire Chief's Car

The author purchased this 1925 Model T in July, 2007 and had the lettering changed to Naval Powder Factory Fire Department. The author began his paid fire career at the former Naval Powder Factory in 1960. PHOTO CREDIT: Bill Killen

1926 Ford Model TT Howe Chemical Pumper

Pittsboro, Indiana's Ford Model TT Howe Model HB-5 Chemical Pumper was equipped with three 70 gallon copper booster tanks, two hose reels and a 150 GPM pump on a one ton chassis. PHOTO CREDIT: Steve Hagy Collection

1926 Bickle Ford Model T 250 GPM Pumper

This 1926 Bickle Ford Model T 250 GPM Combination pumper and hose car was built for the Otterville, Ontario, Canada fire department. PHOTO CREDIT: Walt McCall Collection

1926 Ford Model TT McCarroll

Bob McCarroll's 1926 Ford Model TT before restoration. PHOTO CREDIT: Bob McCarroll

1926 Ford Model TT McCarroll

McCarroll's Ford Model TT was restored with parts of an American LaFrance and is equipped with a chemical tank from a two wheel cart. PHOTO CREDIT: Bob McCarroll

1926 Ford Model TT Luverne before Restoration

"I can't tell you the history because I'm not sure where it served. I was able to determine it was a Luverne because of the steamer cap and the wood used to make the frame of the hose bed had Luverne Minnesota stenciled on it. I determined it was a late model year 1925 because it had a 1926 carb which went into production late in 1925. I was able to locate the great grand daughter of the two brothers who started the Luverne Fire Apparatus Company and sent me some helpful information on the Luverne Model T's."—Andy McConnell PHOTO CREDIT: Andy McConnell

1926 Ford Model TT Luverne after Partial Restoration

1926 Ford Model T Lorne 150 IGPM Pumper

This unrestored rig, lettered for Baden, in Wilmot Township, Ontario, is equipped with a 150 Imperial Gallons per minute pump (IGPM) and 50 Imperial gallon tank. Pump and body was built by Lorne Tractor / Fire Engine Company, Ingersoll Ontario. Purchased by Baden Fire Department it has been in Wilmot Township since. Owned by the Baden Firefighters Association, it was first used July 31, 1927 at the Canadian National Railroad station fire and last used October 17, 1942 at the Shirk and Snider Flour Mill fire. PHOTO CREDIT: Dean Nickerson September 2002.

1926 Ford Model T Gauld Pumper

This 1926 Ford Model T Gauld pumper served the Brigden, Ontario Canada fire department. The front mount pump is either a 250 GPM or 350 GPM. CREDIT: Herb Galloway

1926 Ford Model T Lorne 150 Chemical Truck

Tavistock, Ontario, Canada fire department purchased this double tank chemical pumper October 25, 1926 where it remained in service from 1926 to 1950. The South Easthope Mutual Insurance company restored it 1989 and resides in the South Easthope Mutual Insurance Company Museum. PHOTO CREDIT: Blain Bechtold

1926 Ford Model T Howe Advertisement

This 1926 Howe advertisement features the Opp, Alabama fire department's Howe Model HBS pumping chemical and hose combination PHOTO CREDIT: Steve Hagy Collection.

1926 Ford Model T Howe Pumping Chemical & Hose Combination

Delivery photograph of Opp, Alabama fire department's Howe Model HBS pumping chemical and hose combination PHOTO CREDIT: Steve Hagy Collection.

1926 Ford Model T American LaFrance

This 1926 Ford Model T Fire Truck, with American LaFrance equipment, was the first motorized unit used in Milan, Michigan. The 4-Cylinder 176.7 cubic inch displacement engine produced 20 Horsepower and originally cost $3,050. While visiting the Ford company soybean extraction plant in Milan, Michigan during 1938, Henry Ford noticed this Model T fire truck. Wanting to display it in his museum, he offered to give Milan a new fire engine in exchange for the older truck. This offer was quickly accepted. The names painted on the side of this truck are those of the firemen who were with the Milan Fire Department in 1938. PHOTO CREDIT: Walt McCall Collection

1926 Ford Model T Howe

The Se-Wy-Co Fire Company at Bethlehem, Pennsylvania, operated this Ford Model TT Howe. PHOTO CREDIT: Steve Hagy Collection

Weymouth Nova Scotia's 1926 Ford Model T "Village Queen" Pumper

On December 1, 1924, the Village of Weymouth, Nova Scotia, Canada agreed to enter into a contract with the American LaFrance Fire Engine Co. to purchase the Village's first fire engine a "Village Queen complete with solid rubber tires also 2 – 5 gallon Hand Pumps for $1,580 F.O.B. Weymouth." PHOTO CREDIT: Matt Lee

1926 Ford Model T Horse Drawn Pumper

This unique photograph of an American LaFrance Village Queen Fire Engine in a Yarmouth, Nova Scotia museum was posted on the Model T Ford Club of America's Forum. The fire engine was pulled by hand and the Model T engine served the single purpose of running the fire pump.

1926 Ford Model T Darley Front Mount Pumper

W.S. Darley Company restored this 1926 Darley Front Mount pumper equipped with one of their famous Champion fire pumps. PHOTO CREDIT: W.S. Darley Company.

1926 Ford Model T Darley Front Mount Champion Pump

This 1926 Ford Darley is believed to be one of the earliest Darley front mount pumps on a Model T chassis is owned by Joe & Margaret Arnold, Carlisle, PA. PHOTO CREDIT: Jim Golden

1927 Ford Model T Howe

The Orestes (Indiana) Fire department's 1927 Ford Model T Howe was delivered on April 21, 1927 without a hose bed. It was equipped with twin booster tanks and hose on the reel. PHOTO CREDIT: Steve Hagy Collection

1927 Ford Model TT Howe

Factory photograph of the Trufant, Michigan Fire department's Ford Model TT Howe. PHOTO CREDIT: Steve Hagy Collection

1927 Model T Roadster Pickup Fire Patrol

The Fire Patrol at the Bethlehem Steel Plant in Steelton, Pennsylvania (near Harrisburg) operated this 1927 Roadster Pickup. PHOTO CREDIT: Ken Buohl Collection

1927 Ford Model T American LaFrance

This 1927 Ford Model T one ton chassis and an American LaFrance single tank combination chemical and hose body. PHOTO CREDIT: Phil Gilson Collection

1927 Boyer Ford Model T Double Tank Chemical

At the time of this photograph Jim Golden, Bowie, Maryland owned this 1927 Ford Model T which was originally delivered to the Lille, Maine fire department. PHOTO CREDIT: Tom Herman Collection

SPECIFICATIONS

DELUGE FORD CHEMICAL CAR

Motor and Chassis—Standard Model T Ford complete, less top and deck, with full equipment of Ford tools, etc.

Tires—Standard make, Non-Skid rear, Plain front.

Chemical Tanks—Two (2) steel, leaded inside and out, capacity thirty-gallons each, stationary, operated independently of each other.

Acid Receptacles—Special heavy lead receptacles with patented loose fitting lead cap, held in tinned copper cage. One (1) extra receptacle, cap and cage.

Chemical Extinguishers—One (1) three-gallon nickel plated F. D. type, with improved holder, mounted on right running board. One (1) 1-quart extinguisher mounted on cowl dash above steering post.

Hose Reel—Automatic, loose winding, capacity 200 feet of ¾ inch chemical hose. Will not leak. Hose can be wound up or taken from reel without kinking or interfering with the flow of chemical stream, while in operation. Reel is light but of solid and durable construction.

Hose—150 ft. of ¾ in. 5-ply Deluge special red-covered chemical hose, with heavy improved brass couplings.

Nozzle—Polished brass, special shut-off.

Agitator—Long flat blade extending nearly full length of tank, fastened to shaft, operated by crank, thoroughly disolves soda and mixes contents.

Discharge Pipe—Large extra heavy pipe, arranged to drain tank nearly dry. Directly connected to automatic hose reel.

Valves—Heavy brass, angle shut-off valves.

Drain Valve—Brass valve in front end of tank, with extension handle to operate from rear of machine.

Torches—Two (2) N. P. torches for stationary lighting.

Search Light—One (1) hand electric portable "Grether" fire department search light, mounted on right side of car.

Lantern—One (1) regulation fire department lantern mounted in spring holder.

Warning Signal—Sirene horn, fire department style, also regular Ford magneto horn.

Axe—One (1) fireman's pick-head axe, with holder.

Crowbar—One (1) 42 in. steel crowbar, with holders.

Buckets—Ten (10) metal buckets, painted; for filling tanks.

Funnel—One (1) large round filling funnel.

Ladder—One (1) solid side, 22 ft. Windlass extension fire ladder, two sections, mounted on left side of car in substantial brackets, held with quick release pin, no straps or buckles.

Ladder Brackets—Made of heavy tire steel, securely mounted to car.

Hook Pole—One (1) 10 ft. hook pole with steel hooks, mounted in spring holders, attached to ladder brackets.

Tool Box—One (1) tool and charge box mounted on left running board for carrying extra charges, tools, etc.

Tools—Full equipment of Ford tools, also large monkey wrench.

Chemical Charges—Four (4) complete double charges soda and acid.

Springs—Extra main leaf fit in regular Ford spring, also auxiliary coil springs attached providing ample load carrying capacity.

Towing Attachment—A strong device attached to rear of chassis so that any other apparatus may be towed behind car and taken to fire quickly.

Weight—Car complete with all equipment, 2050 pounds.

Lettering—As ordered.

Finish—Chassis, body and fenders standard black enamel striped in red or white; chemical equipment and wheels grey or red at option.

Extras—Electric starter, demountable rims, special tires, roof ladder, by-pass or other extra equipment furnished at additional cost on special order.

Prospect Fire Engine Company Specification Deluge Ford Chemical Car

Prospect Fire Engine Company's specifications (circa 1927) for their Deluge Ford Chemical Car built on a "Standard Model T Ford complete, less top and deck, with full equipment of Ford Tools, etc." PHOTO CREDIT: Tod Wirth Collection

ABOUT THE AUTHOR

The author began his fire service career at the age of 16 as a volunteer firefighter with the Potomac Heights Volunteer Fire Department, Indian Head, Maryland and in June 1960 he was appointed a probationary Firefighter with the Naval Ordnance Station Fire Department, Indian Head, Maryland.

His fire service career spans fifty-two years and includes service with the Kennedy Space Center Fire Department where he served on the Apollo and Skylab Astronaut Rescue Teams. He served as the Director, Navy Fire and Emergency Services 1985-2004 and Fire Chief, Holston Army Ammunition Plant, Kingsport, Tennessee 2005-2006.

The author served as President of the International Association of Fire Chiefs 2005-2006, and the IAFC Board of Directors 2003-2007. He currently serves as Vice-President National Fire Heritage Center (Heritage Hall of America's Fire Services and Fire Protection Disciplines, Inc), represents the International Association of Fire Chiefs on the board of the International Fire Service Research Center and Policy Institute and is a Past President of the International Association of Fire Chiefs Foundation and currently serves on their board of directors.

He has authored several papers for fire service technical publications, "Navy Marine Corps Fire Apparatus 1836-2000 Photo Gallery", and co-authored the history of the Kennedy Space Center Fire Department and a contributing author for the book "Firefighters".

He is a fire apparatus history buff and restored a 1923 Stoughton 350 GPM Rotary Gear fire engine and is currently writing the history of Stoughton Wagon Works. He is a member of the Model T Ford Club of America, Model T Ford Club International, and Society for the Preservation and Appreciation of Motorized Fire Apparatus in America, the Antique Automobile Club of America, Antique Truck Historical Society and a member of the Bristol VA/TN Fire Museum board of directors.

Bill and his wife of nearly fifty years, Carole, reside in Church Hill, Tennessee.